# MICROELECTRONIC PACKAGING

## A Bibliography

# IFI DATA BASE LIBRARY

---

**COMPUTER TECHNOLOGY**
Logic, Memory, and Microprocessors — A Bibliography
*A. H. Agajanian*

**ACOUSTIC EMISSION**
A Bibliography with Abstracts
*Thomas F. Drouillard*

**MICROELECTRONIC PACKAGING**
A Bibliography
*A. H. Agajanian*

**ENVIRONMENTAL IMPACT STATEMENT GLOSSARY**
A Reference Source for EIS Writers, Reviewers, and Citizens
*Marc Landy*

**SPEECH COMMUNICATION AND THEATER ARTS**
A Classified Bibliography of Theses and Dissertations, 1973—1978
*Merilyn Merenda and James Polichak*

# MICROELECTRONIC PACKAGING

## A Bibliography

Compiled by

## A.H. Agajanian

*IBM Data Systems Division*
*Hopewell Junction, New York*

IFI/PLENUM • NEW YORK-WASHINGTON-LONDON

Library of Congress Cataloging in Publication Data

Agajanian, A  H

  Microelectronic packaging.

  Includes index.
  1. Microelectronic packaging — Bibliography. II. Title.
Z5838.M5A34        [TK7874]              016.381'7                79-18930
ISBN 0-306-65183-1

©1979 IFI/Plenum Data Company
A Division of Plenum Publishing Corporation
227 West 17th Street, New York, N.Y. 10011

Printed in the United States of America

To the memory of my parents

# Preface

Microelectronic packaging protects and supports electronic devices and circuits and provides connections to the other parts of the system. The protection function avoids mechanical, electrical, chemical, contamination, and photo-optical damage, degeneration, and causes of malfunction. Hybrid microelectronic circuits and subsystem packages support the substrates; the substrate contains the circuit elements, (semiconductor devices or IC chips, deposited or chip resistors and capacitors, and attached inductors), as well as deposited and bonded interconnection wires. The connections to other parts of the system include electrical leads, heat removal paths, and mounting functions. At present, in order to meet the demands of VLSI, the emphasis is on packages with higher densities while maintaining performance, reliability, and low cost.

This book is a comprehensive bibliography of over 3000 references of the world literature in microelectronic packaging. It is compiled to assist the workers in the field in comparing their work with that done by others. For easy access to the needed references, the book is divided into many sections and subsections (see Contents). A comprehensive subject index is also given to assure easy access to the needed data. The book cites a number of books and review articles for the beginner in the field who wishes to become familiar with the subject. Novel technologies, such as bubble domain and multilayer ceramic packaging are highlighted.

The literature from January 1976 to December 1978 is covered. The literature prior to 1976 is the subject in part of a previous book by the author.[1] The main sources searched for references are: Electrical and Electronics Abstracts (EEA), Chemical Abstracts (CA), Ceramic Abstracts, and Engineering Index. In addition to these sources, current journals, current conference digests, Books in Print, and Cumulative Book Index are searched for material. The volumes and numbers of the abstracts are given for access to the abstracts.

The author would like to thank Mr. R. B. Murphy, IBM East Fishkill Library manager, for encouragement and support, my wife Else for encouragement and assistance, my colleagues Dr. L. V. Gregor and Dr. J. J. Gniewek for many helpful discussions, Mr. L. A. Plaushin for editorial assistance and the IBM East Fishkill Library staff for their valuable assistance. A special thank you to Mrs. F. Sisson of Graphic Techniques for typing the manuscript.

AVAILABILITY OF DOCUMENTS

The references in this book consist of technical articles pre-
sented in journals, books, symposia proceedings, technical meetings,
reports, patents and theses.  If a reader does not have access to a
library to obtain copies of the references we suggest the following
sources:

1.  Journal articles, books, symposia proceedings:
            Library of Congress
            Photoduplication Service
            Washington, D. C. 20540

            Engineering Societies Library
            345E 47th Street
            New York, New York 10017

2.  U. S. government contract reports:
            National Technical Information Service
            Springfield, Virginia 22151

3.  IAEC reports
            International Atomic Energy Agency
            Kaerntnerring A 1010
            Vienna, Austria

            National Lending Library
            Boston Spa, England

4.  Theses listed in Dissertation Abstracts + number (U.S.)
            University Microfilms
            Dissertation Copies
            P.O. Box 1764
            Ann Arbor, Michigan 48106

     Others
            University Microfilms, Ltd.
            St. John's Road
            Penn, Buckinghamshire
            England

5.  Patents:  U. S. and foreign
            Patent Office
            Box 9
            Washington, D. C. 20231

# Contents

I.   BIBLIOGRAPHIES

1. Agajanian, A. H., "Semiconducting devices: A bibliography of
   fabrication technology, properties, and applications", IFI/
   Plenum, New York, 1976, p. 831-921, EEA80-36297.
2. Agajanian, A. H., "Computer technology: A bibliography of logic,
   memory, and microprocessors", IFI/Plenum, New York, 1978.
3. Hench, L. L. and McEldowney, B. A., (Eds.), "A bibliography of
   ceramics and glass", 2nd Edition, The American Ceramic Society,
   65 Ceramic Drive, Columbus, Ohio, 1976.
4. Reed, W. E., "Diffusion bonding: A bibliography with abstracts",
   Report NTIS/PS-75/771/6SL, 1975, 147 pp.
5. Reed, W. E., (Ed.), "Hybrid microelectronics circuits: Citations
   from the Engineering Index data base, Vol. 1, 1973-74", Report
   NTIS/PS-77/0021/0EES, 1977, 240 pp.
6. Reed, W. E., (Ed.), "Hybrid microelectronic circuits: Citations
   from the Engineering Index data base, Vol. 2, 1975-Feb. 1977",
   Report NTIS/PS-77/0222/8EES, 1977, 150 pp.
7. Reed, W. E., (Ed.), "Hybrid microelectronic circuits: Citations
   from the NTIS data base, 1964-Feb. 1977", Report NTIS/PS-77/
   0220/2ENS, 1977, 198 pp.
8. Reed, W. E. and Lehmann, E. J., "Seals and gaskets: A bibliography with abstracts", Report NTIS/PS-75/838/3SL, 1975, 229 pp.
9. Taylor, C. H., "Plastic encapsulated semiconductor devices – a
   bibliography [1973-1976]", Microelectron. & Reliab., $\underline{16}$ (6),
   701-4 (1977), EEA81-10294.

II.   BOOKS

10. Allen, B. M., "Soldering and welding", Drake Pubs., Inc., New
    York, 1975.
11. Anon., "Packaging with plastics", Gordon & Breach, New York,
    1975.
12. Beiser, L. and Marshall, G. F., (Eds.), "Laser scanning compo-
    nents and techniques – design considerations/trends", SPIE
    Seminar Proceedings, Vol. 84; SPIE Palos Verdes Estate, CA, 1977.
13. Briston, J. H., "Plastic films", Wiley, New York, 1974.
14. Dezettel, L. M., "Electrical soldering", H. W. Sams, Indiana-
    polis, IN, 1976.
15. Glaser, A. and Sabak-Sharpe, G. E., "Integrated circuit engi-
    neering fabrication, and design", Addison-Wesley Publishing
    Co., Reading, Mass., 1977.
16. Gutcho, M. H., "Microcapsules and microencapsulation tech-
    niques", Noyes Data Corp., Park Ridge, N. J., 1976.
17. Hallmark, C. L., "Microelectronics", Tab Books, Blue Ridge
    Summit, Pa., 1976.
18. Hyman, R., "Ceramics handbook", Arco Publishing Co., Inc., New
    York, 1977.
19. Jewett, C. E., "The engineering of microelectronic thin and
    thick films", Macmillan Press, Riverside, N. J., 1978.

20. Jones, E. B., "Electronic components handbook", Reston Publishing Co., Inc., Reston, Virginia, 1977.
21. Lenk, J. D., "Handbook of integrated circuits: for engineers and technicians", Reston Publishing Co., Reston, Va., 1978.
22. Levy, S. and DuBois, H., "Plastics product design engineering handbook", Van Nostrand Reinhold Co., New York, 1977.
23. Martin, L. F., "Fluxes, solders, and welding alloys", Noyes Data Corp., Park Ridge, N. J., 1973.
24. McGonnagle, W. J., (Ed.), "International advances in non-destructive testing, Vol. 5", Gordon & Breach, New York, 1977.
25. Meiksin, Z. H., "Thin and thick films for hybrid microelectronics", Lexington Books, Lexington, Mass., 1976.
26. Mittal, K. L., (Ed.), "Adhesion measurement of thin films, thick films, and bulk coating", American Society for Testing and Materials, Philadelphia, Pa., 1978.
27. Moschytz, G. S., "Linear integrated networks: design", Van Nostrand Reinhold Co., New York, 1975.
28. Pampuch, R., "Ceramic materials: an introduction to their properties", Elsevier Scientific Publications Co., New York, 1976.
29. Ramney, M. W., "Silicones, Vol. 1: Rubber, electrical moulding resins and functional fluids", Noyes Data Corp., Park Ridge, N. J., 1977, EEA81-9441.
30. Ramney, M. W., "Epoxy resins and products: recent advances", Noyes Data Corp., Park Ridge, N. J., 1978.
31. Roddy, D., "Introduction to microelectronics", Pergamon Press, New York, 1978.
32. Rosenthal, M. P., "Mini-micro soldering and wire wrapping", Hayden Book Co., Inc., Rochelle Park, N. J., 1978.
33. Rothenberg, G. B., "Glass technology - recent developments", Noyes Data Corp., Park Ridge, N. J., 1976.
34. Sacharow, S., "Handbook of packaging materials", AVI Publishing Co., Inc., Westport, CT, 1976.
35. Scott, A. W., "Cooling of electronic equipment", Wiley, New York, 1974.
36. Shields, J., "Adhesive bonding", Engineering Design Guides, No. 2, Oxford University Press, London, England, 1974.
37. Tallan, N. M., "Electrical conductivity in ceramics and glass", Marcel Dekker, New York, 1974.
38. Thomas, R. H., "Ultrasonics in packaging and plastics fabrication", Cahners Publishing Co., Inc., Boston, MA, 1974.
39. Thwaites, C. J., "Soldering", Oxford University Press, New York, 1975.
40. Towers, T. D., "Hybrid microelectronics", Crane-Russak Co., New York, 1977.
41. Wang, F. F. Y., (Ed.), "Ceramic fabrication processes", Academic Press, Inc., New York, 1976.
42. Wheeler, G. J., "Electronic assembly and fabrication", Reston Publishing Co., Inc., Reston, Virginia, 1976.

III.   REVIEW ARTICLES

43. Amey, D. I. and Moore, R. P., "LSI package standard and its interconnection variations", Proc. Tech. Program Natl. Electron Packag. Prod. Conf., 1977, p. 65-80.

44. Anon., "Interconnection techniques in the electronic industry - a methods guide", Report IVF-Resultat-72606A, Inst. Verkstadsteknisk Forskning, Gothenburg, Sweden, 1974, 41 pp., Swedish, EEA81-23571.

45. Anon., "Thick films: what and why?", Electron, no. 84, 19, 21 (1975), EEA79-8981.

46. Anon., "Packaging for hazardous environments (electronic equipment)", Electron. Prod. Methods & Equip., 4 (9), 46, 50 (1975), EEA79-11899.

47. Anon., "Potting and encapsulation: technology update", Circuits Manuf., 17 (8), 46-8, 50-1 (1977), EEA81-18362.

48. Anon., "Hybrids vs. monolithics", Mil. Electron/Countermeas., 3 (9), 68-9 (1977), EEA81-19518.

49. Anon., "ISHM report: advances in thick film technology", Circuits Manuf., 17 (12), 42-3 (1977).

50. Anon., "Automation outlook for semiconductor fabrication", Circuits Manuf., 18 (9), 28-30, 34, 36 (1978).

51. Avery, L., "Recent advances in LCI processing and packaging technology", Microelectron. Reliab., 15 (1), 75-83 (1976).

52. Bailey, J., "Component packaging", Eng. Mater. & Des., 22 (6), 45-50 (1978), EEA81-41412.

53. Bauer, C. L. and Lessmann, G. G., "Metal-joining methods", In: Annual review of materials science vol. 6, Huggins, R. A. (Ed.), Annual Reviews, Palo Alto, Calif., p. 361-87 (1976), EEA79-44940.

54. Bernard, W. J., "Developments in electrolytic capacitors", J. Electrochem. Soc., 124 (12), 403C-9C (1977).

55. Boss, I., "The quiet enclosure revolution", New Electron., 10 (24), 71 (1977), EEA81-22852.

56. Boswell, D. and Campbell, D. S., "Economics of thick and thin film hybrid production in Europe", Electrocompon. Sci. & Technol., 4 (3-4), 219-23 (1977).

57. Bouchard, J. G., "Hybrid technology - best supporting actor", IEEE Trans. Manuf. Technol., MFT-6 (4), 65-8 (1977), EEA81-14341.

58. Brooks, P., "Designer's guide to physical design and packaging. III", EDN, 21 (20), 75-80 (1976), EEA80-12538.

59. Bud, P. J., "General review of soldering as a joining method", Insul./Circuits, 23 (3), 48-57 (1977), EEA80-27982.

60. Buzan, F. E., Grier, J. D., Sr., Bertsch, B. E. and Thoryk, H., "Base metal thick film system: resistors, conductors, and dielectrics", Circuits Manuf., 17 (10), 34-9 (1977).

61. Campbell, D. S., "Progress in hybrid microelectronics in Europe as seen at the first European hybrid microelectronics conference", Proceedings of the 1977 International Microelectronics Symposium, p. 291-3.

62. Chu, T. Y., "A general review of mass soldering methods", Insul./Circuits, 22 (12), 73-5 (1976), EEA80-8588.
63. Cichetti, M. and Roggia, D., "Thin-film technology", Telettra, no. 27, 45-50 (1975), EEA79-28649.
64. Corkhill, J. R., "Thick films at high frequencies", In: Handb. Thick Film Technol., Holmes, P. J. and Loasby, R. C., (Eds.), Electrochem. Publications Ltd.: Ayr, Scot., p. 199-227 (1976), CA88-97819.
65. Corkhill, J. R., "Packaging of hybrid microcircuits in plastics", In: Handb. Thick Film Technol., Holmes, P. J. and Loasby, R. G., (Eds.), Electrochem. Publications Ltd.: Ayr, Scot., p. 333-79 (1976), CA88-106321.
66. Crossland, W. A. and Hailes, L., "Substrates [for thick film technology]", In: Handb. Thick Film Technol., Holmes, P. J. and Loasby, R. C., (Eds.), Electrochem. Publications Ltd.: Ayr, Scot., p. 74-96 (1976), CA88-97813.
67. Dance, B., "Integrated circuit package coding", Electron. Ind., 4 (3), 37 (1978), EEA81-22842.
68. Dobrovolskiy, A. G., "Development of slip moulding methods", Ceramurgia Int., 3 (4), 159-64 (1977).
69. Fisher, H. D., "Circuit devices and assembly", In: Handb. Thick Film Technol., Holmes, P. J. and Loasby, R. G., (Eds.), Electrochem. Publications Ltd.: Ayr, Scot., p. 274-332 (1976), CA88-97821.
70. Friedel, W. and Pottharst, J., "Integrated thin and thick film circuits", NTG-Fachber., 54, 85-96 (1975), German, EEA80-501.
71. Ghate, P. B., Blair, J. C. and Fuller, C. R., "Metallization in microelectronics", Thin Solid Films, 45 (1), 69-84 (1977), EEA81-1221.
72. Gray, P. J., "A review of printed circuit technology", Electron, no. 118, 40, 43 (1977), EEA81-10002.
73. Hamer, D. W., "Thick film technology: III", Ceram. Ind., 106 (2), 22-3 (1976).
74. Hurlimann, W., "Electronic experimental systems - up-to-date", Tech. Rundsch., 69 (9), 3, 5, 7 (1977), German, EEA80-32238.
75. Inokuchi, S. and Suzuki, Y., "Application techniques of integrated circuits - I, II, III, IV, V", Syst. & Control, 20 (8), 410-19 (1976); ibid: 20 (9), 472-9 (1976); ibid: 20 (10), 547-56 (1976); ibid: 20 (11), 621-9 (1976); ibid: 20 (12), 665-74 (1976), Japanese.
76. Jarvinen, E. T. K., "Hybrid microelectronics in Finland", Proceedings of the 1977 International Microelectronics Symposium, p. 294-6.
77. Jennings, C. W., "Review of filament growths and corrosion effects caused by moisture films and DC bias between microcircuit conductors", Report SAND-77-0007, Sandia Labs., Albuquerque, New Mex., 1977, 25 pp.
78. Johansen, I., Arntzen, R., Bergmann, S., Husa, S., Ingebrigtsen, K. A., Johannessen, J. S. and Sandved, J. S., "Examples of thin film applications in electronic devices", Thin Solid Films, 50, 171-85 (1978), EEA81-42152.

79. Jolly, J., "Comparison between thick and thin films technolo-
    gies", International Conference on Manufacturing and Packaging
    Techniques for Hybrid Circuits, 1976, p. 25-33, French,
    EEA79-46558.
80. Kirby, P. L., "Hybrid technology applied to the chip intercon-
    nection problem", Microelectron. & Reliab., 14 (4), 369 (1975),
    EEA79-13054.
81. Knox, R. M., "Dielectric waveguide microwave integrated cir-
    cuits - an overview", IEEE Trans. Microwave Theory & Tech.,
    MTT-24 (11), 806-14 (1976), EEA80-505.
82. Labs, J., "Chip-joining - a review", Nachrichtentech. Elektron.,
    28 (7), 284-7 (1978), German, EEA81-45774.
83. Lane, C. H., "Packages and film resistors for hybrid microcir-
    cuits", 13th Annual Proceedings of Reliability Physics Symposi-
    um, 1976, p. 230-41, EEA79-46577.
84. Lane, G., "New materials developments", Electron. & Microelec-
    tron. Ind., no. 209, 50-1 (1975), EEA79-8989.
85. Loasby, R. G. and Barlow, H., "Special purpose materials and
    processes [in thick film technology]", In: Handb. Thick Film
    Technol., Holmes, P. J. and Loasby, R. G., (Eds.), Electro-
    chem. Publications Ltd.: Ayr, Scot., p. 185-98 (1976), CA88-97818.
86. Lutsch, A. G. K., "Manufacture of microcircuits in a small in-
    dustrialised country", Trans. S. Afr. Inst. Electr. Eng., 67
    (9), 258-73 (1976), EEA80-3293.
87. Lyman, J., "Advances in materials, components, processes, ensure
    hybrid prosperity in the LSI age", Electronics, 49 (15), 92-104
    (1976), EEA79-37710.
88. Lyman, J., "Packaging technology responds to the demand for
    higher densities", Electronics, 51 (20), 117-25 (1978),
    EEA82-649.
89. Mackenzie, J. D., "New applications of glass", J. Non-Cryst.
    Solids, 26 (1-3), 456-81 (1977), EEA81-394.
90. McBride, D. G., "Interconnections for microelectronics", ASME
    Paper No. 76-DE-14, 1976, 5 pp.
91. McKinney, B. L. and Faust, C. L., "Progress in electrode-
    position and related processes - 1952-1977", J. Electrochem.
    Soc., 124 (11), 379C-86C (1977).
92. McMillian, P. W., "Advances in technology and applications of
    glasses", Phys. Chem. Glasses, 17 (5), 193-204 (1976).
93. Markstein, H. W., "Interconnection update", Electron. Packag.
    Prod., 15 (10), 29-32, 34 (1975).
94. Markstein, H. W., "Packaging techniques in consumer electron-
    ics", Electron Packag. & Prod., 18 (4), 47-49, 52-53 (1978).
95. Martin, A. N., "Glasses for the electronics industry", Aust.
    Electron. Eng., 8 (6), 27-30 (1975), EEA79-997.
96. Moore, J., "A solder reflow review", Insul./Circuits, 24 (3),
    49-52 (1978), EEA81-33003.
97. Nutt, G. W., "Sub-rack standardisation: The European dimension",
    Electron, no. 124, 55-6, 59 (1977), EEA81-18363.

98. Oldham, W. G., "The fabrication of microelectronic circuits",
Sci. Am., <u>237</u> (3), 111-14, 119-24, 126, 128, 130 (1977),
EEA81-10382.

99. Oswald, R. G. and de Miranda, W. R., "Application of tape
automated bonding technology to hybrid microcircuits", Solid
State Technol., <u>20</u> (3), 33-8 (1977), EEA80-24629.

100. Pfeifer, H. J. and Wetzko, M., "Layer hybrid technology - a
modern development trend of microelectronics die", Technik,
<u>32</u> (5), 253-6 (1977), German, EEA81-5637.

101. Prokop, J. S. and Williams, D. W., "Chip carriers as a means
of high-density packaging", IEEE Trans. Components, Hybrids
& Manuf. Technol., CHMT-<u>1</u> (3), 297-304 (1978), EEA81-49514.

102. Provance, J. D., "Performance review of thick film materials",
Insul./Circuits, <u>23</u> (4), 33-6 (1977), EEA80-24612.

103. Rapp, J. E., "Recent contributions of ceramic technology to
silicon processing", SME Tech. Pap. Ser. EE, Pap. EE76-563,
1976, 10 pp.

104. Ravenhill, J. and Sunda, J., "Progress in aluminium electrolytic capacitors", New Electron., <u>10</u> (10), 123-4, 127 (1977),
EEA81-5546.

105. Reich, B., "Military IC standardization, acquisition and technology", Proceedings of the 1975 Annual Reliability and Maintainability Symposium, p. 404-7, EEA79-46394.

106. **Roikh, I. L., Koltunova, L. N. and Lebedinskii**, O. V., "Protective coatings produced by ion plating in a vacuum (review)",
Prot. Met., <u>13</u> (6), 545-55 (1977).

107. Ryerson, C. M., "Reliability testing and screening - a general
review paper", Electro-Technol., <u>19</u> (4), 59-71 (1975),
EEA79-27513.

108. Savage, J., "Conductor materials", In: Handb. Thick Film
Technol., Holmes, P. J. and Loasby, R. G., (Eds.), Electro-
chem. Publications Ltd.: Ayr, Scot., p. 97-113 (1976),
CA88-97814.

109. Schnable, G. L., Kern, W. and Comizzoli, R. B., "Passivation
coatings on silicon devices", J. Electrochem. Soc., <u>122</u> (8),
1092-1103 (1975).

110. Seaman, J., "Old friends in new packages: The passive DIP",
Electron. Prod., <u>20</u> (1), 25-32 (1977).

111. Secaze, G., "Automatic electronic testing equipment", Mes.
Regul. Autom., <u>42</u> (5), 35-42 (1977), French, EEA80-37964.

112. Simons, A., "Choosing thick film hybrid microelectronics",
Aust. Electron. Eng., <u>8</u> (5), 19-24 (1975), EEA79-13048.

113. Snigier, P., "Hardware and interconnect devices - keys to
today's high-density packaging", EDN, <u>21</u> (22), 192-4 (1976),
EEA80-24004.

114. Stoeckhert, K., "Recent developments in the field of plastics
packaging", Kunstst Ger. Plast., <u>68</u> (5), 29-30 (1978).

115. Swyt, D. A., "Upcoming dimensional standard aimed at needs of
microelectronics industry", Dimensions NBS, <u>61</u> (12), 22
(1977), EEA81-42135.

116. Sydow, R., "Packaging of semiconductor components", Radio Mentor Electron., <u>44</u> (3), 96-9, 109 (1978), German, EEA81-38141.
117. Traeger, R. K., "Organics used in microelectronics: a review of outgassing materials and effects", 27th Electronic Components Conference, 1977, p. 408-20, EEA80-35903.
118. Walton, B., "Survey of hybrid microelectronics in the UK", Microelectron. & Reliab., <u>15</u> (4), 323-8 (1976), EEA79-46548.
119. Wang, P., "An IC factory in the new age - an overview", Electrochemical Society Spring Meeting, Extended Abstracts, 1977, p. 592-3, EEA81-19525.
120. Whitelaw, D., "A comparison of thick and thin films", Electron. Eng., <u>50</u> (602), 65-7 (1978), EEA81-19163.
121. Wild, H., "Vacuum techniques. I.", Elektrotechnik, <u>28</u> (5), 57-9 (1977), German, EEA80-40319.
122. Williams, J. H., "The military system manufacturer's view on microcircuit standardization", Electron. Packag. & Prod., <u>17</u> (4), 133-6 (1977), EEA81-10029.
123. Wilson, G. C., "A review of accelerated ageing (environmental testing) of printed boards with respect to solderability", Circuit World, <u>4</u> (3), 39-44 (1978), EEA82-615.
124. Zaleckas, V. J., "Laser applications in microelectronics", Conference on Laser and Electrooptical Systems, Digest of Technical Papers, 1976, p. 8, EEA80-9192.

IV.  BONDING TECHNIQUES
     1.  General

125. Adelman, N. T., "Bond stresses in circular piezoceramic benders", Acustica, <u>39</u> (1), 43-6 (1977).
126. Anon., "The heavier the wire the harder it is to bond", Circuits Manuf., <u>15</u> (10), 44 (1975), EEA79-13058.
127. Anon., "Mechanical seals: Selection", Power, <u>120</u> (10), 82-3 (1976), EEA80-8593.
128. Anon., "Advantages of wire wrapping techniques", Electron. Equip. News, 5-6 (1977), EEA81-22838.
129. Becher, P. F., "Fritted thick film conductor adherence: role of firing atmosphere", J. Mater. Sci., <u>13</u> (2), 457-9 (1978), EEA81-14326.
130. Bharti, P. L. and Sarin, S. K., "Grounding of MIC's substrates", J. Inst. Electron. & Telecommun. Eng., <u>21</u> (6), 351-2 (1975), EEA79-13053.
131. Bushmire, D. W., "Solid phase bonding", International Microelectronics Conference, Proceedings of the Technical Program, 1975, p. 39-47.
132. Chookazian, M., "Electromagnetic bonding of thermoplastics - a preview", SPE, National Technical Conference, Proceedings, Plastics in Appliances, 1975, p. 150-4.
133. Hutchins, G. L., "Technique to deposit adhering but etchable metallurgy diamond surfaces", IBM Tech. Disclosure Bull., <u>20</u> (1), 426 (1977), EEA81-13292.

134. Keizer, A., "Mass outerlead bonding + applications to open packages", Proc. of the Tech. Program, Int. Microelectron Conf., 1977, p. 37–46.
135. Miller, F. R., "Advanced joining processes", SAMPE Q, 8 (1), 46–54 (1976).
136. Moreton, J., "Fume hazards in welding, brazing and soldering", Met. Constr., 9 (1), 33–4 (1977).
137. Olesen, F. C., "Integrated circuits on kapton film", Elektronik, no. 9, 8–9 (1976), Danish, EEA79-46413.
138. Peel, M., "Chip carrier sockets - what are the options", Proc. Tech. Program. Natl. Electron Packag. Prod. Conf., 1977, p. 87–97.
139. Rifkin, A. A., Schiller, J. M. and Turetzky, M. N., "Sealant for semiconductor package", IBM Tech. Disclosure Bull., 20 (9), 3446 (1978), EEA82-112.
140. Schmidtke, H., "Solderless joints", Elektroniker, 15 (5), EL28-31 (1976), German, EEA79-31900.
141. Schmidtke, H., "Solderless connections. III", Elektroniker, 15 (7), EL33-7 (1976), German, EEA79-40779.
142. Singh, A., "Wireless bonding techniques", J. Inst. Eng. Electron. & Telecommun. Eng. Div., 56, 85–7 (1976), EEA80-3314.
143. Smith, J. M. and Stuhlbarg, S. M., "Hybrid microcircuit tape chip carrier materials/processing tradeoffs", 27th Electronic Components Conference, 1977, p. 34–47, EEA80-35945.
144. Szezepanski, Z., "Perspective methods in wire-less bonding of semiconductor monolithic integrated circuits", Elektronika, 18 (2), 57–61 (1977), Polish, EEA80-20585.
145. Watkins, W. S., "Grounding and bonding electrical equipment", Plant Eng., 30 (15), 116–19 (1976).
146. Williams, J. D. and Crossland, B., "Solid-phase welding processes - 2. Diffusion bonding", Chart Mech. Eng., 23 (4), 73–5 (1976).
147. Wyatt, R., "Wirewrap v. solder", Electron. Equip. News, 7 (1977), EEA81-22839.

2.  Adhesive

148. Allman, D. J., "Theory for elastic stresses in adhesive bonded lap joints", Q. J. Mech. Appl. Math., 30 (4), 415–36 (1977).
149. Althouse, L. P., "Methods for surface treating metals, ceramics, and plastics before adhesive bonding", Report UCID-16997, California Univ., Livermore, Lawrence Livermore Lab., 1976, 21 pp.
150. Amijima, S., Fujii, T. and Yoshida, A., "Two dimensional stress analysis on adhesive bonded joints", Proc. Jpn. Congr. Mater. Res. 20th, 1976, p. 275–81, Publ. 1977.
151. Amijima, S., Fujii, T., Yoshida, A. and Amino, H., "Dynamic response of adhesive bonded joints", Proc. Jpn. Congr. Mater. Res., 20th, 1976, p. 110–17, Publ. 1977.

152. Anon., "Adhesive bonding aluminum body sheet", Automot. Eng., 84 (3), 24-7 (1976).
153. Arnold, D. B., "Bonding development of improved adhesives for acoustic structures", National SAMPE Technical Conference, Vol. 7, 1975, p. 98-117.
154. Brassell, G. W. and Fancher, D. R., "Electrically insulative adhesives for hybrid microelectronic fabrication", IEEE Trans. Components, Hybrids & Manuf. Technol., CHMT-1 (2), 192-7 (1978), EEA81-33028.
155. Brockmann, W., "Effect of surface preparation on the properties of metal bonding", Metall. 31 (3), 245-51 (1977), German.
156. Carrido, J. J., Gerstenberg, D. and Berry, R. W., "Effect of angle of incidence during deposition on Ti-Pd-Au conductor film adhesion", Thin Solid Films, 41 (1), 87-103 (1977), EEA80-20581.
157. Crane, L. W. and Hamermesh, C. L., "Surface treatment of cured epoxy graphite composites to improve adhesive bonding", SAMPLE J, 12 (2), 6-9 (1976).
158. Dietrick, M. I., Emmert, D. E. and Lederer, D. A., "Optimization of wire adhesion by varying concentration of bonding", J. Elastomers Plast., 9 (1), 77-93 (1977).
159. Fredberg, M. L., "Bonding material for planar electronic device", Report AD-D001893/7SL, Department of the Navy, Washington, D. C., 1974, 7 pp.
160. Fredericks, E. C., "Resole polymeric adhesion promoters", IBM Tech. Disclosure Bull., 20 (7), 2704 (1977).
161. Fritzen, J. S., Wereta, A., Jr. and Arvay, E. A., "Cure monitoring techniques for adhesive bonding processes", Natl. SAMPE Symp. Exhib., Vol. 22, 1977, p. 430-4.
162. Gould, F. T., "Method for obtaining adhesion of multilayer thin films", Patent USA 3986944, Publ. October 1976.
163. Graham, J. A., "Structural adhesive bonding", Mach. Des., 48 (23), 118-23 (1976).
164. Herfert, R. E., "Fundamental investigation of anodic oxide films on aluminum alloys as a surface preparation for adhesive bonding", Report AD-A038068/3SL, Northrop Corp., Hawthorne, Calif., 1976, 111 pp.
165. Hitch, T. T. and Bube, K. R., "Basic adhesion mechanisms in thick and thin films", Report AD-A041959/8SL, RCA Labs., Princeton, N. J., 1977, 29 pp.
166. Jennings, C. W., "Two techniques for characterizing surfaces for adhesive bonding", Adhes. Age., 20, 29-35 (1977), CA88-53975.
167. Kaelble, D. H., "Surface energetics criteria for bonding and fracture", SAMPE Q, 7 (3), 30-3 (1976).
168. Levi, D. W., "Durability of adhesive bonds to aluminum", Appl. Polym. Symp. No. 32: Durability of Adhes. Bonded Struct., 1976, p. 189-99, Publ. 1977.
169. McMillan, J. C., Quinlivan, J. T. and Davis, R. A., "Phosphoric acid anodizing of aluminum for structural adhesive bonding", SAMPE Q, 7 (3), 13-18 (1976).

170. McNally, J. P. and Ronan, C. R., "Metal-to-metal adhesive bonding", Weld. Res. Counc. Bull., no. 220, 22 (1976).
171. Minford, J. D., "Durability of structural adhesive bonded aluminum joints", Adhes. Age., 21 (3), 17-23 (1978).
172. Patterson, R. F., "Adhesive bonding of thermoplastics to metal for non-structural applications", Soc. Plast. Eng. Tech. Pap., 24, 697-9 (1978), CA89-75999.
173. Petrie, E. M., "Guide to successful adhesive bonding, I, II, III", Assem. Eng., 19 (6,7,8), 36-41, 18-22, 26-29 (1976).
174. Richardson, R. W., "Bonding of plastics", Eng. Mater. Des., 21 (5), 60-2 (1977).
175. Russell, W. J. and Garnis, E. A., "Study of the FPL etching process used for preparing aluminum surfaces for adhesive bonding", SAMPE Q, 7 (3), 5-12 (1976).
176. Schrader, M. E. and Cardamone, J. A., "Adhesion promoters for bonding titanium", Report AD-A038549/2SL, David W. Taylor Naval Ship Research and Development Center, Annapolis, Md., 1977, 11 pp.
177. Sharma, S. P., "Some notes on the physics of contact adhesion", Insul./Circuits, 23 (11), 41-5 (1977).
178. Smith, T., "Mechanisms of surface degradation of aluminum after the standard FPL etch for adhesive bonding", Appl. Polym. Symp. No. 32: Durability of Adhes. Bonded Struct., 1976, p. 11-36, Publ. 1977.
179. Vohwinkel, F., "Adhesive bonding of titanium and titanium alloys", Adaesion, 21, 68, 70-2 (1977), German, CA87-24278.
180. Vychodil, M. and Myska, L., "Adhesive for the temporary bonding of optical elements", Patent Czech 171341, Publ. February 1978, CA89-90696.
181. Weirauch, D. F., "Mechanical adhesion between a vitreous coating and an iron-nickel alloy", Am. Ceram. Soc. Bull., 57 (4), 420-23 (1978).
182. Westerdahl, C. A. L., Hall, J. R. and Levi, D. W., "Effect of gas plasma treatment time on adhesive bond strength of polymers", Am. Chem. Soc. Div. Polym. Chem. Prepr., 19 (2), 538-43 (1978).
183. Yamakawa, S., "Hot-melt adhesive bonding of polyethylene with ethylene copolymers", Polym. Eng. Sci., 16 (6), 411-18 (1976).
184. Yamakawa, S., "Adhesive bonding of polyethylene. I. Surface treatments", Nippon Setchaku Kyokaishi, 13 (6), 211-18 (1977), CA88-153693.

### 3.    Beam-Lead

185. Bachman, A. K., "Forming metallization patterns on beam lead semiconductor devices", Patent USA 4011144, Publ. March 1977, CA86-149651.
186. Bendure, A. O. and Piper, W. A., "Face-up bonding of beam lead devices", Report BDX-613-1462, Bendix Corp., Kansas City, Mo., 1976, 35 pp.

187. Brown, U. C. and Sim, J. R., "Failure mechanisms in beam lead semiconductors", IEEE Trans. Parts, Hybrids & Packag., PHP-13 (3), 225-9 (1977), EEA81-908.

188. Buhanan, D., McMillan, F. and Wise, J., "Manufacturing methods and technology program for beam lead sealed junction semiconductor devices", Report AD-A028938/9SL, Motorola, Inc., Phoenix, Ariz., 1976, 52 pp.

189. Bushmire, D. W., Chavez, E. L. and Finley, M. H., "Effects of substrate temperature on silicon nitride cracking of beam lead devices during thermocompression wobble bonding", Report SAND-76-0098, Sandia Labs., Albuquerque, N. Mex., 1976, 20 pp.

190. Coughlin, J. B., Hughes, J. B., Watts, R. S. and Siegert, F. W., "A high speed ECL multiplexer in beam lead, hybrid technology", Electrocompon. Sci. & Technol., 4 (3-4), 185-91 (1977), EEA81-14337.

191. Dais, J. L., "The mechanics of gold beam leads during thermocompression bonding", IEEE Trans. Parts, Hybrids & Packag., PHP-12 (13), 241-50 (1976), EEA80-3313.

192. Dais, J. L. and Howland, F. L., "Fatigue failure of encapsulated gold-beam lead and TAB devices", IEEE Trans. Components, Hybrids & Manuf. Technol., CHMT-1 (2), 158-66 (1978), EEA81-33026.

193. Dawes, C. J., "An evaluation of techniques for bonding beam-lead devices to gold thick films", Solid-State Technol., 19 (3), 23-8 (1976), EEA79-28709.

194. Greig, W. J., Moneika, P. J. and Brown, R. A., "High-density high-reliability multichip hybrid packaging with thin films and beam leads", Proceedings of the 28th Electronic Components Conference, 1978, p. 146-50, EEA81-37923.

195. Gruszka, R. F., "Preloading compliant bonding tape with beam-lead ICs", Solid-State Technol., 20 (9), 61-5 (1977).

196. Hall, R. D., "Method for making beam leads", Report AD-D001948/9SL, Department of the Navy, Washington, D.C., 1975, 8 pp.

197. Harman, G. G., "The use of acoustic emission in a test for beam-lead, TAB, and hybrid chip capacitor bond integrity", IEEE Trans. Parts, Hybrids & Packag., PHP-13 (2), 116-27 (1977), EEA80-32247.

198. Hickman, H. H., Kalke, A. R. and Lach, W., "Computer aided manufacturing in beam lead device bonding", 26th Electronic Components Conference, 1976, p. 61-5, EEA79-37721.

199. Jaccodine, R. J. and Nigh, H. E., "Technique for releasing beam leads from silicon substrate", Tech. Dig., no. 49, 17 (1978), EEA81-28186.

200. Jaffe, D., "Encapsulation of integrated circuits containing beam leaded devices with a silicone RTV dispersion", IEEE Trans. Parts, Hybrids & Packag., PHP-12 (13), 182-7 (1976), EEA80-506.

201. Johansson, J., "Beam lead bonding investigation (thick film)", Report N76-30584/6SL, Svenska Radio A.B., Stockholm, Sweden, 1975, 64 pp.

202. Johansson, J., "Comparison of beam lead bonded devices to thin and thick film", Report N76-31551/4SL, Svenska Radio A.B., Stockholm, Sweden, 1975, 33 pp.

203. Lasch, K. B., Schnaitter, W. N. and Ilgenfritz, R., "Reliability assessment of beam-lead sealed-junction integrated circuits in polymer sealed packages", 27th Electronic Components Conference, 1977, p. 429-37, EEA80-35953.

204. Le Cain, M., "The 'beam-lead': a controversial but attractive formula", Electron. & Microelectron. Ind., no. 217, 28-30 (1976), French, EEA79-28704.

205. Lindberg, F. A., "Method for making beam leads for ceramic substrates", Patent USA 4022641, Publ. May 1977, CA87-15014.

206. Ludwig, D. P., Coucoulas, A., Fossi, R. L., Gruszka, R. F. and Zahurak, L. P., "Compliant bonding with color-anodized aluminum compliant tape", Tech. Dig., no. 43, 29-30 (1976), EEA80-510.

207. McMaster, T. F., Carlson, E. R. and Schneider, M. V., "Subharmonically pumped millimeter-wave mixers built with notch-front and beam-lead diodes", 1977 International Microwave Symposium Digest, p. 389-92, EEA81-5235.

208. Ondrik, M. A., Anderson, J. W. and Potts, E. G., "A new method for testing the adherence of anchor pads on beam lead devices", Insul./Circuits, 23 (4), 39-41 (1977), EEA80-24621.

209. Paulson, R., Barney, J., Leftwich, R. F. and Friedman, R., "Infrared imaging for thermal analysis of quality control and reliability problems", SPIE Seminar Proceedings, Vol. 60, 1975, p. 85-90.

210. Peltier, A., Crabtree, M., Zobel, D. and Wise, J., "Manufacturing methods and technology program for beam lead sealed junction semiconductor devices", Report AD-A023846/9SL, Motorola, Inc., Phoenix, Ariz., 1975, 83 pp.

211. Pestie, J. P. and Fourrier, J. Y., "Design and manufacture of a microwave low-noise transistor having beam-leads", IEEE Trans. Electron Devices, ED-24 (2), 73-9 (1977), EEA80-6641.

212. Pritchard, E. J., "Resin and method for protecting the contacts of beam-lead integrated circuits", Tech. Dig., no. 39, 29-30 (1975), EEA79-1108.

213. Robinson, L. A., "Carrier system for testing and conditioning of beam lead devices", 27th Electronic Components Conference, 1977, p. 1-9, EEA80-35942.

214. Rothbauer, M., "GaAs Schottky-barrier beam-lead diodes", Elektrotech. Cas., 29 (2), 148-51 (1978), EEA81-28456.

215. Ryden, W. D. and Labuda, E. F., "Metallization providing two levels of interconnect for beam-leaded silicon integrated circuits", IEEE J. Solid State Circuits, SC-12 (4), 376-82 (1977), CA87-110406.

216. Smith, L. W. and Wilson, B. G., "Automatic testing and handling of beam-lead transistors and diodes", SME Technical Paper Series EM, 1976, Paper EM76-982, 9 pp.

217. Soos, N. A. and Jaffe, D., "Encapsulation of large beam leaded devices", Proceedings of the 28th Electronic Components Conference, 1978, p. 213-16, EEA81-37936.
218. Spano, J., "New failure analysis techniques for beam lead and multi-level metal integrated circuits", 14th Annual Proceedings Reliability Physics, 1976, p. 279-82, EEA80-17049.
219. Swafford, J. H., "Failure modes of beam-lead semiconductors in thin-film hybrid microcircuits", IEEE Trans. Parts, Hybrids & Packag., PHP-12 (4), 298-304 (1976), EEA80-13288.
220. Youngberg, D. A. and Lenhardt, B. W., "Rework technique removes beam-lead devices intact for postmortem tests", Insul./ Circuits, 21 (12), 21-2 (1975), EEA79-9034.

### 4.  Chip Joining

221. Acello, S., "Mini-pak: a cost-effective leadless chip carrier", Electron. Packag. & Prod., 17 (6), 78-82 (1977).
222. Ahn, J., Chiou, C., Duke, P. and Montelbano, T., "Permalloy solder barrier for devices with gold conductor metallization", IBM Tech. Disclosure Bull., 20 (12), 5317 (1978).
223. Angelone, P. A., "Method for removing flip chips", IBM Tech. Disclosure Bull., 19 (7), 2477 (1976), EEA80-24620.
224. Angelucci, T., Sr., "Gang lead bonding integrated circuits", Solid State Technol., 19 (7), 21-5 (1976), EEA79-37744.
225. Angelucci, T. L., Sr., "Gang lead bonding equipment, materials and technology", Solid State Technol., 21 (3), 65-8 (1978), EEA81-33042.
226. Anon., "Super-8 film for mounting IC chips", Radio Elektron. Schau, 52 (10), 28 (1976), German, EEA80-6427.
227. Anon., "The transfer machine 'TAB'", Electron. & Appl. Ind., no. 235, 39-42 (1977), French, EEA81-5640.
228. Antonucci, R. F., Lennon, W. R., Schiller, J. M. and Schlesier, R., "Prefluxing semiconductor chips", IBM Tech. Disclosure Bull., 21 (6), 2426 (1978).
229. Antonucci, R. F., Lennon, W. and Schlesier, R. M., "Dendritic solder joints", IBM Tech. Disclosure Bull., 21 (7), 2910-11 (1978).
230. Balde, J. W. and Amey, D. I., "New chip carrier package concepts", Computer, 10 (12), 58-68 (1977), EEA81-18360.
231. Barnard, D. J., Bogner, T. and Stowe, E. D., "Module cap ledge reform die", IBM Tech. Disclosure Bull., 20 (2), 527 (1977), EEA81-13302.
232. Bauer, J. A., "Use of chip carriers for high packaging density, high reliability, high performance products", Proc. Tech. Program Natl. Electron Packag. Prod. Conf., 1977, p. 50-6.
233. Bernadotte, C., "Development of the TAB-process at Saab-Scania", Proceedings of the 28th Electronic Components Conference, 1978, p. 151-8, EEA81-37933.

234. Breuninger, K., Haberland, D. R. and Herberger, R., "Crossover chips for hybrid-integrated film circuits", Feinwerktech. & Messtech., 85 (8), 369-70 (1977), German, EEA81-10049.

235. Brunner, J., Jaspan, J. S., Mackey, R. E., Rose, H. K. and Shah, A. S., "Joining chip to substrate in oxygen-containing atmosphere", IBM Tech. Disclosure Bull., 20 (6), 2318 (1977).

236. Brusic, V. and Change, D. A., "Solder ball C4 joints for multireflow", IBM Tech. Disclosure Bull., 21 (4), 1688 (1978).

237. Buchoff, L. S., "Stretching the chip bonding alternatives with reliable, low cost elastomeric connectors", Insul./Circuits, 21 (12), 23-4 (1975), EEA79-9035.

238. Burch, M. L. and Hargis, B. M., "Ceramic chip carrier - the new standard in packaging?", Proc. Tech. Program Natl. Electron Packag. Prod. Conf., 1977, p. 43-9.

239. Burns, C. D. and Kanz, J. W., "Gang bonding to standard aluminized integrated circuits with bumped interconnect tape", Solid State Technol., 21 (9), 79-81 (1978), EEA82-654.

240. Cain, R. L., "Beam tape carriers - a design guide", Solid State Technol., 21 (3), 53-8 (1978), EEA81-33040.

241. Caruso, S. V. and Honeycutt, J. O., "Investigation of discrete component chip mounting technology for hybrid microelectronic circuits", Report NASA-TM-X-64937, NASA, Huntsville, Ala., 1975, 60 pp., EEA79-4785.

242. Cavanaugh, D. M., "Thermal comparison of flip-chip relative to chip-and-wire semiconductor attachment in hybrid circuits: an experimental approach", IEEE Trans. Parts, Hybrids & Packag., PHP-12 (4), 293-8 (1976), EEA80-13287.

243. Chabra, K. C., Gupta, D. K. and Arora, O. P., "Flip chip bonding", J. Inst. Electron. & Telecommun. Eng., 21 (5), 293-3 (1975), EEA79-9038.

244. Coombs, V. D. and Larnerd, J. D., "Circuit chip support assembly with fatigue resistant epoxy bond", IBM Tech. Disclosure Bull., 21 (1), 99-100 (1978).

245. David, R. F. S., "Advances in epoxy die-attach", Solid State Technol., 18 (9), 40-4 (1975), EEA79-1107.

246. DeMuzio, D. D., Griesemer, H. A. and Large, D. M., "Soldering of leads to semiconductor chips", Tech. Dig., no. 44, 15-16 (1976), EEA80-13296.

247. DiGiacomo, G. and Ordonez, J. E., "Use of ternary and quaternary pastes in chip joining", IBM Tech. Disclosure Bull., 20 (7), 2749 (1977), EEA81-45773.

248. Eifert, R. and Maslov, A., "Deformations and internal stresses at the bonding of solid state circuits", Feingeraete Tech., 25 (5), 198-200 (1976), German, EEA79-33163.

249. Fritz, G. F., "Tape assembled components - a packaging option", 16th Annual Proceedings, Reliability Physics Symposium, 1978, p. 124-6, EEA81-48755.

250. Goldmann, L. S., Herdzik, R. D., Koopman, N. G. and Marcotte, V. C., "Lead-indium for controlled-collapse chip joining", IEEE Trans. Parts, Hybrids & Packag., PHP-13 (3), 194-8 (1977), EEA81-917.
251. Gray, H. F., "Some basic problems with eutectic preform material for large wafer bonding", Electrochemical Society Spring Meeting, Extended Abstracts, 1977, p. 51-2, EEA81-5642.
252. Greer, S. E., "Low expansivity organic substrate for flip-chip bonding", Proceedings of the 28th Electronic Components Conference, 1978, p. 166-71, EEA81-37935.
253. Guzenski, A. L., "Method of mounting wafers", Tech. Dig., no. 44, 21-2 (1976), EEA80-13447.
254. Herdzik, R. and Koopman, N. G., "Dummy pads for increased creep resistance", IBM Tech. Disclosure Bull., 20 (4), 1394 (1977), EEA81-23646.
255. Hetherington, D. R., "Bond chips with conductive epoxies", Electron. Des., 23 (20), 82-5 (1975), EEA79-13447.
256. Kamei, T., Nakamura, M., Ariyoshi, H. and Doken, M., "Hybrid IC structures using solder reflow technology", Proceedings of the 28th Electronic Components Conference, 1978, p. 172-82, EEA81-37924.
257. Knapp, E. C. and Roush, W. B., "Solder flux", IBM Tech. Disclosure Bull., 20 (8), 3089 (1978).
258. Koopman, N., "Solder well cap and process", IBM Tech. Disclosure Bull., 20 (3), 995-6 (1977), EEA81-22833.
259. Kuist, C. H., "Connecting LSI chips", 9th Annual Connector Symposium, 1976, p. 351-6, CCA12-10618.
260. Kuntzleman, H. C., Legge, F. V. and Wasson, R. L., "Chip/cable field replaceable unit", IBM Tech. Disclosure Bull., 20 (7), 2619-20 (1977), EEA81-45771.
261. Lassus, M., "Metal bumps aid semiconductor and hybrid circuit producers", Electron. Engineering, 48 (583), 63-7 (1976), EEA79-46544.
262. Lever, R. F., "Applying radiant heat to semiconductor integrated circuits", IBM Tech. Disclosure Bull., 20 (10), 3908-9 (1978).
263. Ludwig, D. P., "Chip-in-tape - its role in hybrid integrated circuit manufacture", Proc. of the Tech. Program, Int. Microelectron. Conf., 1977, p. 22-8.
264. McBride, D. G., "Mechanical chip attachment", IBM Tech. Disclosure Bull., 20 (7), 2615 (1977), EEA81-45769.
265. Martin, B. D., "Individually flip chip bonding semiconductor devices to a substrate", IBM Tech. Disclosure Bull., 21 (5), 1854-5 (1978).
266. Minetti, R. H. and Strickland, R. W., "Method and apparatus for removal of semiconductor chips from hybrid circuits", Patent USA 3969813, Publ. July 1976.
267. Pienkowska, B. and Chrobak, P., "Organic adhesives for die bonding of semiconductors to substrates of hybrid circuits", Elektronika, 17 (10), 363-5 (1976), Polish, EEA80-17042.

268. Rivenburgh, D. L. and Van Vestrout, V. D., "Backbonding an
     integrated circuit to a heat sink", IBM Tech. Disclosure Bull.,
     19 (10), 3692 (1977).
269. Rodrigues de Miranda, W. R., Oswald, R. G. and Brown, D.,
     "Lead forming and outer lead bond pattern design for tape-
     bonded hybrids", IEEE Trans. Components, Hybrids & Manuf.
     Technol., CHMT-1 (4), 377-83 (1978).
270. Rose, A. S., Scheline, F. E. and Sikina, T. V., "Metallurgical
     considerations for beam tape assembly", Solid State Technol.,
     21 (3), 49-52, 68 (1978).
271. Seeba, M. D. and Sears, R. A., "Ceramic-chip-capacitor attach-
     ment", IEEE Trans. Parts, Hybrids & Packag., PHP-13 (4),
     395-9 (1977).
272. Singh, A., "When and why to use passive chips in hybrid cir-
     cuits", Microelectron. & Reliab., 16 (6), 705-6 (1977),
     EEA81-10055.
273. Smith, J. M., "Hybrid microcircuit tape chip carrier materials/
     processing trade offs", IEEE Trans. Parts, Hybrids & Packag.,
     PHP-13 (3), 257-68 (1977), EEA81-910.
274. Soffa, A., "Remarks on wire and die bonding for hybrid cir-
     cuits", Electrocomponent Sci. Technol., 4 (3-4), 157-61 (1977).
275. Szczepanski, Z., "Perspective methods in wire-less bonding of
     semiconductor monolithic integrated circuits", Elektronika, 18
     (2), 57-61 (1977), Polish, EEA80-20585.
276. Tan, S. I., "Individual chip joining monitor", IBM Tech. Dis-
     closure Bull., 21 (6), 2551-2 (1978).
277. Wadhwa, S. K., "Direct chip attach", Proceedings of the 13th
     Electrical/Electronics Insulation Conference, 1977, p. 102-4,
     EEA81-14294.
278. Waite, G. C., "Semiconductor chip attachment with small bump
     flip chips", IEEE Trans. Manuf. Technol., MFT-4 (1), 8-13 (1975).
279. Zawicki, L. R., "Semiconductor die attachment", Report BDX-
     613-1809, Bendix Corp., Kansas City, Mo., 1978, 16 pp.,
     CA89-139271.

## 5. Metal-Ceramic Seals

280. Anderson, N. C., "Metal-to-ceramic seals", Patent USA 3918922,
     Publ. November 1975.
281. Bailey, F. P. and Black, K. J., "Gold to alumina solid state
     reaction bonding", J. Mater. Sci., 13 (5), 1045-52 (1978).
282. Becher, P. F. and Murday, J. S., "Thick film adherence
     fracture energy - influence of alumina substrates", J. Mater.
     Sci., 12 (6), 1088-94 (1977).
283. Burgess, J. F., Neugebauer, C. A. and Moore, R. E., "The direct
     bonding of metals to ceramics and application in electronics",
     Electrocompon. Sci. & Technol., 2 (4), 233-40 (1976),
     EEA79-24052.
284. de Bruin, H. J. and Warble, C. E., "Chemical bonding of metals
     to ceramic materials", Patent USA 4050956, Publ. September 1977.

285. Friedel, R. and Ebersberger, H., "Metal-ceramic soldering connection", Patent USA 3950836, Publ. April 1976.
286. General Electric Co., "Alumina-metal joints", Patent UK 1361225, Publ. July 1974.
287. General Electric Co., "Bonding metal to ceramic", Patent UK 1494951, Publ. December 1977.
288. Hamano, Y., "Method for bonding ceramics with metal", Patent USA 3897624, Publ. August 1975.
289. Lucas, J., "Ceramic-metal joint", Patent UK 1326017, Publ. August 1973; Patent UK 1343839, Publ. January 1974.
290. McVey, C. I., "Nb or Ta coated with fired Zr-Mo for metal-ceramic seals", Patent USA 4076898, Publ. February 1978.
291. Neill, D. E., "The direct bonding of copper to ceramic and resulting advantages and applications", International Conference on Manufacturing and Packaging Techniques for Hybrid Circuits, 1976, p. 239-47, EEA79-46574.
292. Pollins, P., "Ceramic-to-metal feedthrus and cable terminations for making connections under adverse environmental conditions", Insul./Circuits, 24 (13), 83-4 (1978).
293. RCA Corp., "Ceramic metal seal", Patent Japan 78-16374, Publ. May 1978, CA89-139383.
294. Siemens A. G., "Metal-ceramic solder joints", Patent UK 1471438, Publ. April 1977.
295. Snow, G. S. and Wilcox, P. D., "Ceramic-to-metal seal", Patent USA 3951327, Publ. April 1976.
296. Sun, Y. S. and Driscoll, J. C., "A new hybrid power technique utilizing a direct copper to ceramic bond", IEEE Trans. Electron Devices, ED-23 (8), 961-7 (1976), EEA79-36696.
297. Syunry, T. S., "Design and development of ceramic-metal seals for vacuum and ultrahigh-vacuum work", Res. Ind., 22 (3), 164-68 (1977).
298. Takashio, H., "Forsterite ceramic-to-titanium metal seals by reactive alloying method", Yogyo Kyokai Shi, 83 (8), 411-16 (1975).
299. Takashio, H., "Alumina ceramic-to-nickel metal seals by reactive alloying method using Ti-Ni-Ag solder", Yogyo Kyokai Shi, 84 (12), 594-99 (1976).
300. United States of America, "Ceramic-to-metal seal", Patent UK 1475483, Publ. June 1977.

## 6. Metal-Glass Seals

301. Becher, P. F. and Newell, W. I., "Adherence-fracture energy of a glass-bonded thick-film conductor: effect of firing conditions", J. Mater. Sci., 12 (1), 90-6 (1977), EEA80-3299.
302. Bondarev, L. F., Kiselev, A. A., Turuntsov, V. V., Skorik, N. S. and Pafomov, Y. E., "Bonding of metal to glass", Patent USSR 563372, Publ. June 1977, CA87-89388.
303. Buckley, R. G., "Understanding glass-to-metal seals", Electron. Packag. & Prod., 18 (10), 74-9 (1978).

304. Buckley, R. G., "Considerations for selecting and evaluating glass-to-metal seals", Insul./Circuits, 24 (12), 19-20 (1978).
305. Hall, G. R., White, J. M. and Krechmery, R. L., "Glass-to-metal seal", Patent USA 4019388, Publ. April 1977.
306. Johnson, A. H., Langston, P. R., Jr. and Sunners, B., "Multi-layer glass-metal package", Patent Canadian 981799, Publ. January 1976.
307. Kutsukake, Y., Nomaki, K. and Nagano, K., "Glass to metal bonding with metal solder "Cerasolzer" technique", 10th International Congress on Glass, Proceedings, Pt. 1, 1974, p. 103-10.
308. Lamson, M. A. and Ramsey, T. H., Jr., "Improving the quality of glass-sealed ceramic DIPs", Electron. Packag. & Prod., 18 (1), 84-6, 91-2, 94, 96 (1978), EEA81-45172.
309. Matheson, R., "Designing glass-to-metal seals", Electron. Packag. & Prod., 17 (7), 197-201 (1977).
310. Podgorski, T. J., "Manufacture of glass-to-metal seals wherein the cleanliness of the process is enhanced and the leak resistance of the resulting seal is maximized", Report N76-21558/1SL, NASA, Langley Research Center, Langley Station, Va., 1976, 9 pp.
311. Reintgen, R. J., "New approach for the use of glass-to-metal seals", International Microelectronics Conference, Proceedings of the Technical Programs, 1975, p. 217-19.
312. Rekhson, S. M. and Mazurin, O. V., "Stress relaxation in glass and glass-to-metal seals", Glass Technol., 18 (1), 7-14 (1977).
313. Zydzik, G. J., van Uitert, L. G., Singh, S. and Kyle, T. R., "Strong adhesion of vacuum-evaporated gold to oxide or glass substrates", Appl. Phys. Lett., 31 (10), 697-9 (1977), EEA81-9432.

## 7.   Soldering

314. Ackroyd, M. L. and MacKay, C. A., "Solders, solderable finishes and reflowed solder coatings", Circuit World, 3 (2), 6-12 (1977), EEA81-4772.
315. Adamson, C., "Soldering the right tools for the job", Electr. Equip., 16 (2), 25, 27 (1977), EEA80-34985.
316. Agarwala, B. N., Dalal, H. M. and Klepeis, S. J., "Creep resistant solder alloys", IBM Tech. Disclosure Bull., 21 (6), 2415 (1978).
317. Aimi, B. R., "T-shaped finger glass dam", IBM Tech. Disclosure Bull., 20 (2), 528-9 (1977), EEA81-14358.
318. Ameen, J. G., Elmore, G. V. and Peter, A. E., "Strippable solder mask material comprising polysulfone silicon dioxide filler, and solvent", Patent USA 4120843, Publ. October 1978.
319. Anon., "Nickel improves Ni-Cr wire solderability", Circuits Manuf., 16 (10), 46-7 (1976), EEA80-3244.
320. Anon., "Common production soldering problems: causes and cures", Circuits Manuf., 17 (7), 96-100 (1977), EEA81-19127.

321. Anon., "Solderability testing of component leads", Report
     Mekanresultat-77006, Sveriges Mekanforbund, Stockholm, Sweden,
     1977, 19 pp., Swedish, English, EEA81-22837.
322. Asahi Glass Co., Ltd., "Method of and apparatus for soldering
     electrical leads to difficultly solderable substrates", Patent
     UK 1409964, Publ. October 1975.
323. Asami, E., Aoyama, H. and Kobayashi, K., "Fluxes for solder-
     ing", Patent Japan 77-97345, Publ. August 1977, CA88-93679.
324. Bakovets, V. V., "Wetting of silicon by lead, tin, and ger-
     manium in hydrogen and hydrogen chloride", Inorg. Mater., 13
     (4), 596-7 (1977), EEA81-23847.
325. Bankes, R. B. and Grubb, J. W., "One step soldering of diode
     bridges in a nest", Tech. Dig., no. 49, 5-6 (1978), EEA81-27203.
326. Bascom, W. D. and Bitner, J. L., "Void reduction in large-area
     bonding of IC components (solder bonding)", Solid State
     Technol., 18 (9), 37-9 (1975), EEA79-1071.
327. Bernard, C., "Wave soldering joint quality trouble shooting
     guide", Insul./Circuits, 23 (12), 23-5 (1977), EEA81-14283.
328. Bernard, C. D., "Horizontal vs inclined conveyor wave-soldering,
     which approach reduces solder bridges more effectively?",
     Circuits Manuf., 17 (9), 53-5 (1977).
329. Bernard, C. D., "An alternative to thermal profiling in wave-
     soldering", Circuits Manuf., 17 (2), 34 (1977), EEA81-5605.
330. Bernier, D., "Soldering hybrid microcircuits with paste solder",
     Insul./Circuits, 21 (12), 27-8 (1975), EEA79-9036.
331. Bernier, D., "Tips on high temperature solder and soldering",
     Insul./Circuits, 23 (10), 41-2 (1977), EEA81-129.
332. Bizinskaya, E. A. and Bakovets, V. V., "Some characteristics
     of the wetting of Ge by tin and lead", Inorg. Mater., 13 (4),
     598-9 (1977), EEA81-23848.
333. Boynton, K. G., "Moisture, fixturing, and cleaning-key ingred-
     ients for successfully wave soldering flexible circuits",
     Insul./Circuits, 23 (9), 43-6 (1977), EEA80-41060.
334. Brown, V. L., "Bonded gold fingers as a low-cost alternative
     to patterned edgeboard fingers for general PWB use", IEEE
     Trans. Components, Hybrids & Manuf. Technol., CHMT-1 (3),
     274-81 (1978), EEA81-49494.
335. Browne, L. T. and Malcolm, I., "Soldering hybrid circuits",
     Electron. Prod. Methods & Equip., 4 (7), 72-4, 77, 79 (1975),
     EEA79-4780.
336. Buchnov, G. M. and Kobikov, S. I., "Experience gained during
     introduction of an installation for group soldering of units",
     Prib. and Sist. Upr., no. 5, 43-4 (1978), Russian, EEA81-44959.
337. Cioffi, J. M. and Moore, P. F., "Pin and solder ring loader",
     IBM Tech. Disclosure Bull., 21 (7), 2912-13 (1978).
338. Condra, L. W., "Vapor condensation soldering of external leads
     to thin-film hybrid integrated circuits", 27th Electronic
     Components Conference, 1977, p. 135-40, EEA80-35962.

339. Davies, D. G., "Soldering to silver-plated semiconductors", Circuits Manuf., 16 (5), 52, 54, 56, 58 (1976), EEA79-36694.

340. Davies, R. L., "High strength low temperature bonding with silver-tin solders", Weld J., 55 (10), 838-42 (1976).

341. Doyle, K. B., "Some experiences and conclusions using soldered and welded packages for hermetic thick film hybrids", Microelectron. & Reliab., 16 (4), 303-7 (1977), EEA81-10053.

342. Dubey, G. C., "Wave soldering of printed circuit boards", J. Inst. Electron Telecommun. Eng., 21 (7), 386-8 (1975).

343. Etler, K., "Solder-resist lacquer", Elektroniker, 16 (6), EL26-9 (1977), German, EEA81-9233.

344. Feddersen, P. A., "Controlling solder microcircuit", Circuits Manuf., 16 (4), 20-3 (1976), EEA79-37741.

345. Fitak, A. G., "Solder: constitution, analysis, and behavior", Electron. Packag. & Prod., 17 (7), 224-6, 228-31 (1977).

346. Friebe, E. R., "Induction soldering evaluation", Report BDX-613-1453, Bendix Corp., Kansas City, Mo., Energy Research and Development Administration, 1976, 45 pp.

347. Frieser, R. G., Powell, J. L. and Tummala, R. R., "Crystallization in copper-containing solder glasses", J. Electrochem. Soc., 125 (3), 492-8 (1978).

348. Fuchs, H., "Economical soldering techniques", Elektro-Anz., 31 (1-2), 28-30 (1978), German, EEA81-32229.

349. Gould, D. J. and Wild, R., "Solder process", IBM Tech. Disclosure Bull., 20 (11), 4721 (1978).

350. Green, M. W. and McKie, G. W., "Wave or flow soldering and hot oil levelling as a quality assurance tool", Electron. Prod., 6 (6), 31-1, 33, 35, 37 (1977), EEA81-14301.

351. Heinzel, H. and Saeger, K. E., "Wettability and dissolution behaviour of gold in soft soldering", Gold Bull., 9 (1), 7-11 (1976).

352. Herdzik, R. J. and Koopman, N. G., "Dual solder evaporation process", IBM Tech. Disclosure Bull., 20 (8), 3087-8 (1978).

353. Herdzik, R. J., Sullivan, E. J. and Totta, P. A., "Timing preplated sites on a substrate", IBM Tech. Disclosure Bull., 19 (8), 3049-50 (1977), EEA80-35955.

354. Hershberger, R. F., Kohn, H. and Senger, R. C., "Immersible solder wave system", IBM Tech. Disclosure Bull., 20 (6), 2156-7 (1977).

355. Hyman, H., "Backplane soldering with preforms", Insul./Circuits, 24 (5), 25-7 (1978), EEA81-45724.

356. Inoue, H., Yasuda, T. and Funyu, I., "Precision soldering", Patent Japanese 75-62155, Publ. May 1975, CA84-172855.

357. Inoue, K., "Automated soldering system operation stage-by-stage", JEE J. Electron. Eng., no. 98, 39-42 (1975).

358. Iseda, T., Nomaki, K. and Nagano, K., "Soldering of semiconductor elements on substrates", Patent Japanese 75-139676, Publ. November 1975, CA84-172857.

359. Jellison, J. L., Johnson, D. R. and Hosking, F. M., "Statistical interpretation of meniscograph solderability tests", IEEE Trans. Parts, Hybrids and Packag., PHP-12 (2), 126-33 (1976), EEA79-37646.

360. Johannessen, J. S., "Is tin soldering a real field of research?", Elektro, 90 (4), 16, 19-21 (1977), Norwegian, EEA80-16452.

361. Kadota, T., Hayasaka, T. and Yoshihara, H., "A study of film-soldering with gold eutectic alloys", J. Vac. Soc. Jap., 19 (2), 46-54 (1976), Japanese, EEA79-37743.

362. Kakhramanov, K., Mazur, V. A., Roshal, R. M. and Akhmedli, G. T., "High-temperature solders for thermoelement connections based on lead chalcogenides", Appl. Sol. Energy, 12 (5), 9-11 (1976), EEA80-23997.

363. Kaska, H., "Surface coating effects on soldering quality in electronic equipment", Wiad. Telekomun., 17 (9), 255-9 (1977), Polish, EEA81-14287.

364. Keller, J. and Keller, J. D., "Temperature controlled soldering irons", Circuits Manuf., 16 (11), 46, 48, 50, 52 (1976), EEA80-16451.

365. Knapp, E. C. and Roush, W. B., "Solder flux", IBM Tech. Disclosure Bull., 20 (9), 3089 (1978).

366. Kobayashi, T., Kamibayashi, T., Okada, A., Mimura, Y. and Mitsumoto, N., "Soldering of ultrafine wires and printed circuits", Patent Japan 77-66856, Publ. June 1977, CA87-210268.

367. Koromyslova, G. I., Selivanov, A. N. and Nosaev, G. A., "Binder for a paste-like solder", Patent USSR 560719, Publ. June 1977, CA87-171692.

368. Kruglov, L. D., "Ultrasonic check on the melting point of solder when soldering small parts", Sov. J. Non-destr. Test., 12 (3), 332-5 (1976).

369. Langan, J. P., "How to decide if an alloy is an adequate substitute for tin-lead solder", Insul./Circuits, 22 (1), 20-1 (1976), EEA79-15892.

370. Langan, J. P., "Procedures that assure high reliability and maximum efficiency in wave soldering equipment", Insul./Circuits, 22 (2), 15-16 (1976), EEA79-24053.

371. Langan, J. P., "Tips on selection of wire solder for manual and automatic soldering", Insul./Circuits, 23 (13), 33-4 (1977), EEA81-18357.

372. Langan, J. P., "How to improve component lead solderability", Insul./Circuits, 24 (9), 33-5 (1978).

373. Langan, J. P., "How to select fluxes for maximum effectiveness", Insul./Circuits, 24 (13), 79-80 (1978).

374. Langan, J. P. and Souzis, L., "Functional alloy approach to soldering", Weld J., 56 (1), 13-17 (1977).

375. Lee, I. W. H., "Screenable solder paste", Weld. J., 56 (10), 32-6 (1977).

376. Lilienthal, P. F., Wenger, G. M. and Zado, F. M., "Residue removal methods for condensation soldering systems", Electron. Packag. & Prod., 18 (4), 175-6, 178, 180-1 (1978).

377. Loeffler, J. R., Jr., "N/C laser soldering - fast, low cost, no rejects", Assem. Eng., 20 (3), 32-4 (1977).
378. Loffel, J., "The solderability of electronic components", Fernmeldetechnik, 17 (4), 131-2 (1977), German, EEA81-9235.
379. McGougan, C., "Soldering aluminium - use the right materials", Des. Eng., 44-5, 46, 48 (1976), EEA79-27571.
380. Maier, W., "The finer points of soldering", Funk-Tech., 32 (22), 399-400, 409-10 (1977), German, EEA81-18347.
381. Manko, H. H., "Tune up your hand soldering!", Assem. Eng., 19 (7), 24-6 (1976).
382. Manko, H. H., "Selecting solder alloys for hybrid bonding", Insul./Circuits, 23 (4), 27-30 (1977), EEA80-28629.
383. Manko, H. H., "How to solder microelectronic assemblies", Weld. Des. & Fabr., 50 (8), 77-8 (1977), EEA81-13299.
384. Manko, H. H., "Understanding of solder wave and its effects on solder joints", Insul./Circuits, 24 (1), 45-9 (1978), EEA81-28159.
385. Mehl, R. M., "Hand soldering, production and in-process repair", SME Tech. Pap. Ser. EE, Paper EE77-921, 1977, 22 pp.
386. Miles, F. D., "Soldering of aluminium and its alloys", International Electronic Packaging and Production Conference, INTERNEPCON UK '73, Proceedings of the Technical Programme, 1973, p. 83-96.
387. Minetti, R. H. and Rickabaugh, L. J., "Solder dissolution rates of evaporated and sputtered Ti.Pd.Au and NiCr.Au thin films", 27th Electronic Components Conference, 1977, p. 212-19, EEA80-35912.
388. Mitsugi, S., Oelschlagel, D., Yamaguchi, K., Yamaji, K. and Konishi, K., "Clad solder for semiconductor devices", Weld. J., 56 (10), 301S-5S (1977).
389. Mohler, J. B., "Soldering properties of metal surfaces, I, II", Plant Eng., 30 (22, 23), 109-10, 153-4 (1976).
390. Moorhead, A. J., Woodhouse, J. J. and Easton, D. S., "Soldering of copper-clad niobium-titanium superconductor composite", Weld. J., 56 (10), 23-31 (1977).
391. Moser, F. R., Murphy, K. J., Palchik, S. S. and Schuessler, P. V., "Water soluble flux for lead tin, lead indium solder joints and stacked joints", IBM Tech. Disclosure Bull., 21 (3), 938 (1978).
392. Nelding, H., "Solderability - a problem when soft soldering", Eltek. Aktuell Elktron. A, 19 (15), 52-3 (1976), Swedish, EEA80-36.
393. Olsen, D. and Wright, R., "Solderability of external lead coatings after storage between 25 and 400°C", Thin Solid Films, 45 (1), 203-4 (1977), EEA81-130.
394. Orletskii, V. B., Kondratenko, M. M. and Tovstyuk, K. D., "Soldering of contacts to $Pb_{1-x}Sn_xTe$ single crystals having a low concentration of current carriers", Instrum. & Exp. Tech., 19 (4), 1223 (1976), EEA80-34988.

395. O'Rourke, H. T., "Controlling wave soldering through the use of time/temperature instrumentation", Insul./Circuits, 24 (11), 27-9 (1978).

396. Owen, C. J. and Poliak, R. M., "Water-soluble flux", IBM Tech. Disclosure Bull., 20 (2), 551-2 (1977), EEA81-19137.

397. Pantanelli, G. P., "Measurement of the solderability of thick-film circuits: relationship between solder strength and a solderability test", Solid State Technol., 18 (10), 39-41 (1975), EEA79-4735.

398. Peterkort, W. G., "Low temperature aluminum soldering analysis", Report BDX-613-1557, Bendix Corp., Kansas City, Mo., 1976, 48 pp.

399. Pietrzykowski, J., "Wave soldering of electronic component. Possibilities of process optimization", Wiad. Telekomun., 16 (7-8), 207-15 (1976), Polish, EEA79-44939.

400. Plegat, A. E., "Solder for joining aluminum to other alloys", Patent USA 4032059, Publ. June 1977, CA87-171693.

401. Rickabaugh, L. J., "The effect of thin film deposition angle and substrate surface roughness on film dissolution in molten 60% tin-40% lead solder", Electrocomponent Sci. Technol., 4 (1), 43-6 (1977), CA89-172547.

402. Rivenburgh, D. L. and Romanosky, R. J., "Cast-solder preloading for stacked modules", IBM Tech. Disclosure Bull., 20 (2), 545-6 (1977), EEA81-19136.

403. Rubin, W., "Oxide-free solder creams", Electron. Prod. Methods & Equip., 5 (3), 17, 19 (1976), EEA79-31902.

404. Ruhl, R. L., Metz, L. D. and Heckman, R., "Reducing the low-level incidence of solder shorts in wave soldering", Electron. Packag. & Prod., 17 (2), 82-5 (1977).

405. Rusignuolo, N., "Nondestructive hand soldering of microelec-tronics", SME Tech. Pap. Ser. EE, Paper EE77-117, 1977, 14 pp.

406. Schoenthaber, D., "Solder coating thickness considerations for hot gas solder leveling", Insul./Circuits, 24 (12), 39-43 (1978).

407. Schuessler, P. W., "Water-soluble flux for lead/indium solder joints", IBM Tech. Disclosure Bull., 20 (9), 3394 (1978).

408. Schuessler, P. W., "Water soluble flux", IBM Tech. Disclosure Bull., 20 (11), 4303 (1978).

409. Seifert, G., "Removing solder from electronic components", Elektromeister & Dtsch. Elektrohandwerk, 52 (10), 705-6 (1977), German, EEA80-31388.

410. Shah, A. S. and Vandersea, J. E., "Improving tensile strength of multiple reflowed solder joints", IBM Tech. Disclosure Bull., 20 (3), 1096 (1977), EEA81-18373.

411. Slater, P. J., "Soldering and point splitting", J. Comb. Theory Ser. B, 24 (3), 338-43 (1978), EEA81-37306.

412. Sleppin, M., "The advantages of soldering flat packs and lead frames with infrared equipment", Insul./Circuits, 23 (8), 35-6 (1977), EEA81-915.

413. Smith, R. L. and Hewett, R. D. "Soldering with active organic-acid flux", Electron. Packag. Prod., 15 (11), 57-58, 60-61 (1975).

414. Spigarelli, D. J., "Vapor phase solder reflow for hybrid cir-
     cuit manufacturing", Solid State Technol., 19 (10), 50-3
     (1975), EEA80-3309.
415. Stelmak, J. P., "Soldering to electroless nickel", IBM Tech.
     Disclosure Bull., 20 (7), 2668 (1977), EEA81-45170.
416. Stoneman, A. M., McKay, C. A., Thwaites, C. J. and Mackowiak,
     J., "Oxidations of molten solder alloys under simulated wave-
     soldering conditions", Met. Technol., 5 (pt. 4), 126-32 (1978).
417. Storchai, E. I., Shalaeva, O. N. and Galkina, T. N., "Thermal
     stability of chloride-fluoride fluxes for the soldering of
     aluminum", Chem. Pet. Eng., 13 (7-8), 720-2 (1977).
418. Teitz, P. D., "Application of solder creams by screen-process
     printing (hybrid ICs)", Microelectronics, 5 (2), 55-6 (1973),
     EEA79-9039.
419. Thomas, G. M., "Solder aging - 10/90 tin lead solder", IBM
     Tech. Disclosure Bull., 20 (2), 549 (1977), EEA81-18351.
420. Thwaites, C. J. and Evans, C. J., "Soft soldering", Engineer-
     ing, 218 (4), I-XI (1978).
421. Turner, R. L., "Troubleshooting the wavesoldering system",
     Circuits Manuf., 18 (11), 24, 26, 28 (1978).
422. Van der Molen, T., "Water soluble fluxes vs. rosin fluxes for
     soft soldering of electronic devices", Insul./Circuits, 23
     (5), 32-8 (1977), EEA80-34989.
423. Walgren, L., "Methods and tools for solder joints removal",
     Insul./Circuits, 22 (9), 13-16 (1976), EEA80-480.
424. Ward, W. C., "Low-temperature soldered component removal",
     IBM Tech. Disclosure Bull., 19 (7), 2476-8 (1976), EEA80-23999.
425. Ward, W. C., "Low-melt solder alloy adds a new dimension to
     soldering technology", Electron. Packag. & Prod., 17 (5),
     98-102 (1977), EEA81-9234.
426. Wenger, G. M., "Meniscograph measurement of the solderability
     retention provided by organic cover coats", Proc. Tech. Pro-
     gram Natl. Electron. Packag. Prod. Conf., 1977, p. 277-84.
427. Wheeler, J., "Solder masking solder plated circuits", Can.
     Electron. Eng., 22 (1), 50-1, 54 (1978), EEA81-32994.
428. Wright, C., "The effect of solid-state reactions upon solder
     lap shear strength", IEEE Trans. Parts, Hybrids & Packag.,
     PHP-13 (3), 202-7 (1977), EEA81-918.
429. Wright, J. J., "Glycolic acid wave soldering flux", IBM Tech.
     Disclosure Bull., 21 (4), 1432 (1978).
430. Zhukov, V. V. and Melik-Ogandzhanyan, P. B., "Investigation of
     the process of soldering of printed-circuit units", Weld
     Prod., 23 (5), 16-18 (1976).

## 8.   Thermocompression

431. Ahmed, N. and Svitak, J. J., "Characterization of gold-gold
     thermocompression bonding", Solid State Technol., 18 (11),
     25-32 (1975), EEA79-9041.

432. Blazek, R. J. and Piper, W. A., "The optimization of lead frame bond parameters for production of reliable thermocompression bonds", Proceedings of the 28th Electronic Components Conference, 1978, p. 373-9, EEA81-42169.

433. Bonham, H. B., "High-reliability thermocompression bonding for hybrid applications", IEEE Trans. Components, Hybrids & Manuf. Technol., CHMT-1 (3), 223-7 (1978), EEA81-49512.

434. Castonguay, R., "What data to consider when selecting fiber glass reinforced plastic substrates for thermal compression bonding", Insul./Circuits, 23 (10), 45-7 (1977), EEA81-916.

435. Condra, L. W., Svitak, J. J. and Pense, A. W., "The high temperature deformation properties of gold and thermocompression bonding", IEEE Trans. Parts, Hybrids & Packag., PHP-11 (4), 290-6 (1975), EEA79-13060.

436. Haddad, M. M., Kowalczyk, R. J. and Suierveld, J., "Technique for forming pads for thermal compression bonding", IBM Tech. Disclosure Bull., 21 (6), 2316 (1978).

437. Ishizaka, A., Iwata, S. and Yamamoto, H., "Formation of clean Al surface by interface deformation in Au-Al thermo-compression bonding", J. Jpn. Inst. Met., 41 (11), 1154-60 (1977), Japanese, EEA81-10063.

438. Iwata, S., Ishizaka, A. and Yamamoto, H., "High-speed thermo-compression bonding of Au wires to Al electrodes on semiconductor devices", J. Jpn. Inst. Met., 41 (11), 1161-5 (1977), Japanese, EEA81-10064.

439. Johnson, D. R. and Willyard, D. L., "The influence of lead frame thickness on the flexure resistance and peel strength of thermocompression bonds", 26th Electronic Components Conference, 1976, p. 80-5, EEA79-37624.

440. Kotani, M., "Design fabrication of low noise Gunn diode with consideration of a thermocompression bonding effect", IEEE Trans. Electron Devices, ED-23 (6), 567-72 (1976).

441. Lach, T. M., "Thermal compression bond failures due to incomplete sintering of alumina ceramics", Proceedings of the 28th Electronic Components Conference, 1978, p. 7-15, EEA81-37932.

442. Laczko, B., Ujvari, A. and David, B., "Measurement of interface resistivity increase - a test method for the degradation of thermocompression bonds", Hiradastechnika, 29 (3), 72-8 (1978), Hungarian, EEA81-49508.

443. Mateev, S. M., Gjurkovski, S. S., Stamenov, K. V. and Tomov, I. V., "Laser technology for the recuperation of bonder needles for thermocompression welding in microelectronics", Elektron Prom-st. & Priborostr., 11 (6), 213-15 (1976), Bulgarian, EEA80-3312.

444. Mykietyn, E. and Veloric, H. S., "A compressible pillar for flip-chip thermocompression bonding", RCA Tech. Not., no. TN 1175, 1-2 (1977), EEA80-35957.

445. Panousis, N. T., "Thermocompression bondability of bare copper leads", IEEE Trans. Components, Hybrids & Manuf. Technol., CHMT-1 (4), 372-7 (1978).

446. Panousis, N. T. and Hall, P. M., "The effects of gold and nickel plating thicknesses on the strength and reliability of thermocompression-bonded external leads", IEEE Trans. Parts, Hybrids & Packag., PHP-12 (4), 282-7 (1976), EEA80-13295.

447. Panousis, N. T. and Hall, P. M., "Thermocompression bonding of copper leads plated with thin gold", 27th Electronic Components Conference, 1977, p. 220-5, EEA80-35968.

448. Panousis, N. T. and Hall, P. M., "Reduced gold-plating on copper leads for thermocompression bonding - Part I: initial characterization, Part II: long term reliability", IEEE Trans. Parts, Hybrids & Packag., PHP-13 (3), 305-13 (1977), EEA81-919.

449. St. Pierre, R. L., Riemer, D. E. and Williamson, M. H., "The 'dirty' thick film gold conductor and its effect on bondability", 26th Electronic Components Conference, 1976, p. 98-102, EEA79-37648.

450. Weiss, B. L., "A method of studying thermocompression bonding damage in GaAs using the SEM", Microelectronics, 7 (1), 35-7 (1975), EEA79-7851.

451. Weiss, B. L. and Hartnagel, H. L., "Localized plastic deformation of GaAs produced by thermocompression bonding", Int. J. Electron., 43 (2), 105-17 (1977), EEA80-36193.

## 9. Ultrasonic

452. Antonevich, J. N., "Fundamentals of ultrasonic soldering", Weld J., 55 (7), 200-07 (1976).

453. Davies, D. G., "Wire bonding to silver plated semiconductor components", Circuits Manuf., 16 (4), 24, 26, 28, 30 (1976), EEA79-36693.

454. Denslow, C. A., "Ultrasonic soldering", Wire J., 9 (9), 131-6 (1976).

455. Giannandrea, C., "What is ultrasonics, and how is it used", SME Tech. Pap. Ser. AD, Paper AD77-722, 1977, 20 pp.

456. Graff, K., "Macrosonics in industry: ultrasonic soldering", Ultrasonics, 15 (2), 75-81 (1977), EEA80-27977.

457. Harman, G. G. and Albers, J., "The ultrasonic welding mechanism as applied to aluminum- and gold-wire bonding in microelectronics", IEEE Trans. Parts, Hybrids & Packag., PHP-13 (4), 406-12 (1977), EEA81-14360.

458. Horowitz, S. J., Gerry, D. J. and Cote, R. E., "Ultrasonic aluminum wire bonding and lead/indium soldering to gold alloy thick film conductors - performance and failure mechanism", AFTA 77, 1977, p. 227-35, EEA81-14363.

459. Horowitz, S. J., Gerry, D. J. and Cote, R. E., "Alloy element additions to gold thick film conductors: effects on indium/lead soldering and ultrasonic aluminium wire bonding", Solid State Technol., 21 (1), 47-54 (1978), EEA81-28185.

460. Hulst, A. P. and Lasance, C., "Ultrasonic bonding of insulated wire", Weld. J., 57 (2), 19-25 (1978).

461. Johnson, K. I., Scott, M. H. and Edson, D. A., "Ultrasonic wire welding. I. Wedge-wedge bonding of aluminium wires", Solid State Technol., 20 (3), 50-6 (1977), EEA80-24630.
462. Johnson, K. I., Scott, M. H. and Edson, D. A., "Ultrasonic wire welding. II. Ball-wedge wire welding", Solid State Technol., 20 (4), 91-5 (1977), EEA80-35958.
463. Khadpe, S. and Bull, D. N., "Ultrasonic bondability of platinum-silver conductors in hybrid microcircuits", 26th Electronic Components Conference, 1976, p. 86-91, EEA79-37722.
464. Kholopov, Y. V., "On the similarity of ultrasonic welding processes", Weld. Prod., 24 (4), 1-3 (1977), EEA81-32234.
465. Koleshko, V. M. and Muzhichenko, O. G., "Effects of ultrasonic microwelding condition parameters on the quality of joints in integrated circuits", Autom. Weld., 28 (4), 43-4 (1975), EEA79-11904.
466. Martin, B. D., "Design and use of a laser interferometer for ultrasonic bonding studies", 14th Annual Proceedings Reliability Physics Symposium, 1976, p. 82-5.
467. Mazur, A. I., Shorshorov, M. K. and Alekhin, V. P., "Oscillation amplitude of the tool in ultrasonic welding", Autom. Weld., 28 (3), 18-19 (1975), EEA79-11902.
468. Mozgovoi, I. V., Samchelkin, V. V., Mozogovoi, V. I., Oshchepkov, V. E., Gonashevskii, L. V., Shestel, L. A., Sokolov, V. A. and Poluyuanov, V. I., "Apparatus for the super-high-speed filming of the ultrasonic welding of plastics", Autom. Weld., 28 (5), 16-17 (1975), EEA79-15890.
469. Parker, R. J., "Ultrasonic interconnection bonding of photovoltaic solar cells", SME Tech. Pap. Ser. AD, Paper AD77-725, 1977, 10 pp.
470. Pfluger, A. R. and Sideris, X. N., "New developments in ultrasonic welding", SAMPE Q, 7 (1), 9-19 (1975).
471. Pietrzykowski, J., "More important applications of ultrasonic vibrations in electronics industry", Wiad. Telekomun., 16 (11-12), 361-9 (1976), Polish, EEA80-24002.
472. Smith, M., "Ultrasonic wire bonding: what you don't know can cost you yield", Circuits Manuf., 17 (10), 18-24 (1977).
473. Winchell, V. H., II, "The mechanism of ultrasonic wire bonding", Electrochemical Society Spring Meeting, Extended Abstracts, 1976, p. 260-2, EEA79-44941.
474. Winchell, V. H., II, "Evaluation of silicon damage resulting from ultrasonic wire bonding", 14th Annual Proceedings Reliability Physics Symposium, 1976, p. 98-107.
475. Winchell, V. H., II, and Berg, H. M., "Enhancing ultrasonic bond development", IEEE Trans. Components, Hybrids & Manuf. Technol., CHMT-1 (3), 211-19 (1978), EEA81-49510.

10.  Welding/Brazing

476. Anderson, D. G., "Laser welding and cutting systems", SME
     Technical Papers Series MR, Paper MR76-856, 1976, 13 pp.
477. Anderson, W. A., "Metallurgical studies of the vacuum brazing
     of aluminum", Weld. J., $\underline{56}$ (10), 314S-18S (1977).
478. Anikin, L. T., Kravetskii, G. A. and Dergunova, V. S., "High
     temperature brazing of graphite using an aluminum brazing
     alloy", Weld. Prod., $\underline{24}$ (7), 23-5 (1977).
479. Anon., "Solid state welding", Proceedings of the 2nd Inter-
     national Symposium of the Japanese Welding Society, 1975,
     p. 209-74.
480. Anon., "Vibration welding permits novel designs", Des. Eng.,
     47-8 (1978), EEA81-13286.
481. Becher, P. F. and Balen, S. A., "Joining of $Si_3N_4$ and SiC ce-
     ramics via solid state brazing", Proc. of the DARPA/NAVSEA
     (Def. Adv. Res. Proj. Agency/Nav. Sea Syst. Command) Ceram.
     Gas Turbine Demonstr. Eng. Program Rev., 1978, p. 649-54.
482. Besednyi, A. V., "Reaction brazing of titanium alloys with a
     copper interlayer thicker than 100 µm", Weld. Prod., $\underline{23}$ (11),
     47-9 (1976).
483. Bowen, B. B., "Preparation of aluminum alloy surface for spot
     weld bonding", National SAMPE Technical Conference Vol. 7,
     1975, p. 374-85.
484. Chirvinskii, S. S., Kovalevskii, R. E., Otmakhova, N. G. and
     Ivakhnenko, I. S., "X-ray cinematographic examination of the
     kinetics of filling capillary gaps during high temperature
     brazing", Weld. Prod., $\underline{23}$ (1), 77-8 (1976), EEA80-2560.
485. DeCristofaro, N. and Henschel, C., "Brazing foil", Weld. J.,
     $\underline{57}$ (7), 33-8 (1978).
486. Doherty, P. E. and Harraden, D. R., "New forms of filler
     materials for diffusion brazing", Weld. J., $\underline{56}$ (10), 37-9 (1977).
487. Engel, S. L., "Laser cutting of thin materials", SME Technical
     Paper MR74-960, 1974, 11 pp.
488. Hellier, C., "Weld inspection - today and tomorrow", Weld.
     Des. & Fabr., $\underline{50}$ (9), 78-81 (1977), EEA81-22874.
489. Herbrich, H., "Under which conditions is welding and cutting
     by means of laser economical?", Gasdynames and Chemical Lasers,
     Proceedings of the International Sympoisum, 1976, p. 157-70.
490. Hetherington, D., "Solderless thick film hybrids", Electron,
     no. 91, 29, 31 (1976), EEA79-20562.
491. Hovey, C. L., "Thin film diffusion brazing of copper-2 Be",
     SME Tech. Pap. Ser. AD, Paper AD77-727, 1977, 11 pp.
492. Hurley, D., "Design potentials of heat-welding plastic parts",
     Plast. Eng., $\underline{34}$ (4), 29-32 (1978).
493. Johnson, D. R. and Knutson, R. E., "Parallel gap welding to
     thick-film metallization", IEEE Trans. Parts, Hybrids & Packag.,
     PHP-$\underline{12}$ (13), 187-94 (1976), EEA80-499.
494. Kawano, H., Oelschlagel, D. and Yamaji, R., "Development of a
     composite brazing wire", Weld. J., $\underline{56}$ (10), 325S-30S (1977).

495. Kirshnaswamy, H. N. and Boccelli, V. E., "Micro-circuit flat-pack sealing by laser welding", SAMPE Q, 8 (4), 11-19 (1977).

496. Kosnac, L., "Physical and chemical properties of the filler materials for brazing and soldering", Weld. Met. Fabr., 44 (4), 294-5 (1976).

497. Kostrubiec, F. and Leszczynski, J., "The welding technology of thin wires by a laser beam", Elektronika, 18 (1), 30-3 (1977), Polish, EEA80-19683.

498. Marciniak, F. R., "How to fixture for better brazing and soldering", Assem. Eng., 21 (7), 34-7 (1978).

499. Miller, C. P., "Resistance welding", Circuits Manuf., 18 (8), 14, 16-18 (1978).

500. Miller, L. F., "Brazing paste", IBM Tech. Disclosure Bull., 21 (1), 137 (1979).

501. Nakagawa, Y., "Aluminum brazing", Patent Japan 77-86948, Publ. July 1977, CA88-54097.

502. Naruki, K. and Hanai, M., "Vacuum brazing", J. Vac. Soc. Jap., 204-11 (1976), Japanese, EEA79-44936.

503. Nicholas, E. D., "Friction welding: a state-of-the-art report", Weld. Des. Fabr., 50 (7), 56-62 (1977).

504. Schaffer, W., "A contribution to the study of welding of contact materials under the influence of bouncing", Elektrie, 30 (11), 595-7 (1976), German, EEA80-5817.

505. Tikhomirov, A. V., "Gas laser cutting conditions for thin sheet materials", Weld. Prod., 22 (5), 15-17 (1975).

506. Tutorskaya, N. N., "Brazing alloys with low vapour tension not containing precious metals", Weld. Prod., 23 (12), 43-6 (1976).

507. Williams, B. R., "Basics of copper brazing", Weld. Des. Fabr., 51 (4), 63-5 (1978).

508. Upite, G., Varcena, S. and Manik, Y. E., "Cold welding of a semiconductor to a metal", Avtom. Svarka, no. 5, 22-5 (1976), Russian, CA85-39853.

## 11.  Others

509. Akimov, V. N., Rydzevskii, A. P., Fedorov, B. I. and Leonov, V. S., "Choosing a calculation scheme for the temperature field in pulse microwelding", J. Eng. Phys., 29 (5), 1450-6 (1975), EEA80-24627.

510. Anon., "Vapor bonding technique", Patent Japan 77-152847, Publ. December 1977, CA89-156492.

511. Ashkinazi, G. A., Galinskii, E. R., Golosov, V. V., Timofeev, V. N., Khutoryanskii, E. D., Chelnokov, V. E. and Yakobson, Y. P., "Rectifying properties of the aluminum-strongly doped gallium arsenide contacts formed by the method of diffusion bonding", Pis'ma Zh. Tekh. Fiz., 4 (10), 596-600 (1978), Russian, CA89-172541.

512. Bassous, E., "Bonding together surfaces coated with silicon dioxide", IBM Tech. Disclosure Bull., 19 (7), 2777-8 (1976).

513. Benjamin, C. E., "High-current internal contact structure for integrated circuits", IBM Tech. Disclosure Bull., 19 (10), 3732 (1977).

514. Brady, M. J., Bogardus, E. H. and Lane, R., "Field induced bonding of optical fibers to glass coated silicon", IBM Tech. Disclosure Bull., 20 (11), 4653 (1978).

515. Burns, J. A. and DiLeo, D. A., "Batch bonded crossovers for thin film circuits. I. Development", Western Electric Eng., 20 (2), 2-10 (1976), EEA79-33096.

516. Chrobak, P. and Fortuna, E., "The coupling of semiconductor elements to substrates in hybrid thick-film microcircuits", Elektronika, 17 (7-8), 265-9 (1976), Polish, EEA80-3303.

517. Crispin, R. M. and Nicholas, M., "Wetting and bonding behaviour of some nickel alloy-alumina systems", J. Mater. Sci., 11 (1), 17-21 (1976).

518. Eggemann, R. V., "Aluminum for bonding Si-Ge alloys to graphite", Patent USA 3931673, Publ. January 1976.

519. Featherby, M., "Gold alloy for Al bonding cuts MOS assembly costs", Circuits Manuf., 17 (6), 68, 72-4, 76-7 (1977), EEA81-14353.

520. Goto, E., "Bonding method using a soldering glass", Patent USA 4066427, Publ. January 1978.

521. Jeglum, T. R., "Capturing components by swage: cutting and flattening leads in one step", Circuits Manuf., 18 (2), 32, 34, 36 (1978), EEA81-49491.

522. Kirkman, D. H., "Bonding glass to polycarbondate", IBM Tech. Disclosure Bull., 20 (11), 4946 (1978).

523. Kniese, W., "Solderless connection technology for backpanels", Elektronik, 27 (7), 55-7 (1978), German, EEA81-42109.

524. Komamoto, O., "Resistance stability of solderless wrapped connection with fine wires", Electr. Commun. Lab. Tech. J., 25 (7), 1081-8 (1976), Japanese, EEA80-3272.

525. Larionov, I. N. and Chernetskii, T. P., "Joining semiconductor single crystals or metals", Patent UK 1328185, Publ. August 1973.

526. Naiki, T. and Seto, S., "Developing research for practical use of diffusion bonding. Diffusion bonding in argon atmosphere", Ishikawajima-Harima Eng. Rev., 17 (4), 354-64 (1977), Japanese, EEA81-4776.

527. Nickols, S. E. and Fay, R. M., "Bonding of piezoelectric materials", IBM Tech. Disclosure Bull., 21 (7), 2986 (1978).

528. Peters, R. D., "Diffusion bonding techniques", Report GEPP-310, General Electric Co., St. Petersburg, Flag., 1978, 16 pp.

529. Ralston, J. R., Jr., "Conductive flexible seal and mounting strip", IBM Tech. Disclosure Bull., 20 (3), 1109-10 (1977), EEA81-27204.

530. Schneider, R. A. and Kroon, W. L., "Magnetic head face assembly using chemically machinable glass-ceramic bonding means", Patent USA 3913143, Publ. October 1975.

531. Sergent, J. E., "Epoxy bonding in hybrid microelectronic cir-
     cuits", International Conference on Manufacturing and Packaging
     Techniques for Hybrid Circuits, 1976, p. 123-31, EEA79-46563.
532. Sletten, E. and Villien, P., "Microwave integrated circuit
     process investigation. Work package 1: Device mounting using
     conductive epoxies", Report N76-31429/3SL, Danish Research
     Center for Applied Electronics, Hoersholm, 1975, 153 pp.
533. Smith, G. C., "Energy pulse bonding – a quick, reliable method
     for terminating flat cable", Insul./Circuits, 22 (1), 29-30
     (1976), EEA79-15893.
534. Varmazis, C., Viswanathan, R. and Caton, R., "Technique for
     bonding gold and silver metals on sapphire", Rev. Sci.
     Instrum., 49 (4), 549-50 (1978), EEA81-33037.
535. Wasserman, B. and Kaufman, M. H., "Process for bonding poly-
     mers", Report AD-D002111/3SL, Department of the Navy, Washing-
     ton, D. C., 1975, 7 pp.
536. Zaderei, N. N., "Research on seals with a leak-proof ring",
     Russ. Eng. J., 57 (8), 37-8 (1977), EEA81-45186.


V.   INTERCONNECTIONS
     1.   Single Layer


537. Alcorn, G. E., Feeley, J. D. and Lyman, J. T., "Process for
     forming a ledge free aluminum copper silicon conductor struc-
     ture", Patent USA 4062720, Publ. December 1977.
538. Anon., "Gold in hybrid micro-electronics", Elektronik, 24 (11),
     103-5 (1975), German, EEA79-9027.
539. Anon., "Advances in platinum-silver thick-film conductors",
     Platinum Met. Rev., 19 (4), 153 (1975).
540. Anon., "Guidelines for the qualification of solderless/mechan-
     ical electrical interconnection techniques", Report N77-10444/
     6SL, European Space Research and Technology Center, Noordwijk,
     Netherlands, 1975, 11 pp.
541. Anon., "Interconnection system suits all electronic needs",
     Des. Eng., 7-9 (1976), EEA79-36683.
542. Archer, J. D., "Fibre optics interconnection components",
     AGARD Conf. Proc. No. 219; Optic Fibres, Integr. Optic and
     Their Mil. Appl., Pap. presented at the Electromagn. Wave
     Propag. Panel/Avionics Panel Jt. Symp., 1977, p. 1-52.
543. Bardo, G. B., "Card terminal block assembly", IBM Tech. Dis-
     closure Bull., 20 (1), 89 (1977), EEA81-14306.
544. Battey, G. L., "Interface signal splitter", IBM Tech. Dis-
     closure Bull., 20 (9), 3637 (1978), EEA82-612.
545. Beck, W. P. and Sim, J. R., "Interconnection technology",
     Report BDX-613-1346, Bendix Corp., Kansas City, Mo., 1976,
     60 pp.
546. Bonham, H. B., "Hybrid microcircuit intraconnection processes",
     Report BDX-613-1257, Bendix Corp., Kansas City, Mo., 1976,
     84 pp.

547. Bowen, T., "Fiber optics as an interconnecting medium", Electron. Packag. Prod., _16_ (4), 17–32 (1976).

548. Burns, T. A., Dunman, J. P. and Smart, A. P., "Programmable flexible tape cable", IBM Tech. Disclosure Bull., _20_ (5), 1932–3 (1977), EEA81–32953.

549. Bushmire, D. W., "Gold aluminum interconnect stability on thin film hybrid microcircuit substrates", 14th Annual Proceedings Reliability Physics Symposium, 1976, p. 55–62, EEA80–17045.

550. Bushmire, D. W., "Resistance increases in gold aluminum interconnects with time and temperature", IEEE Trans. Parts, Hybrids & Packag., PHP–_13_ (2), 152–6 (1977), EEA80–32254.

551. Caley, R. H., "Gold in thick-film conductors", Gold Bull., _9_ (3), 70–5 (1976).

552. Catt, I., Davidson, M. F. and Walton, D. S., "Interconnection of logic elements", Wireless World, _84_ (1510), 61–3 (1978).

553. Cefarelli, F. P. and Evans, R. T., "Three fiber system for optical interconnections", IBM Tech. Disclosure Bull., _20_ (9), 3571–2 (1978).

554. Cefarelli, F. P. and Evans, R. T., "Optical circuit module connector", IBM Tech. Disclosure Bull., _21_ (4), 1568–70 (1978).

555. Crandall, R. W., Freeman, C. D., Flink, H. G. and Frosio, P. J., "Back panel element connector", IBM Tech. Disclosure Bull., _20_ (2), 722–3 (1977), EEA81–14307.

556. Cutillo, J. G., Summa, W. J. and Zucconi, T. D., "High resolution circuitization process", IBM Tech. Disclosure Bull., _20_ (9), 3389–90 (1978), EEA82–617.

557. Doo, V. Y., "Liquid metal multihead connector", IBM Tech. Disclosure Bull., _20_ (11), 4789–90 (1978).

558. Evans, R. T., "Inter-card '0' insertion force connector", IBM Tech. Disclosure Bull., _20_ (6), 2324–6 (1977), EEA82–628.

559. Graube, M., "Stitching circuits together [printed circuit techniques]", Insul./Circuits, _24_ (9), 27–9 (1978), EEA82–623.

560. Horowitz, S. J., Gerry, D. J. and Cote, R. E., "Connecting to gold thick-film conductors", Electron. Packag. & Prod., _17_ (11), 57–68 (1977).

561. Jardine, L. J. and Grossman, S. E., "Interconnection via discrete wiring", Electron. Packag. & Prod., _17_ (10), 39–40, 42, 46, 48–49 (1977).

562. Jarvis, D. C., "Packaging of line replaceable units utilizing rigid-flex-rigid master interconnects", Proceedings of Southeastcon '78 Region 3 Conference, p. 282–4, EEA81–37136.

563. Joshi, K. C., McBride, D. G. and Spaight, R. N., "Multiple I/O pin terminations for packaging", IBM Tech. Disclosure Bull., _20_ (9), 3403–4 (1978), EEA82–110.

564. Liu, T. S., "Aspects of gold-tin bump-lead interconnection metallurgy", Proceedings of the 1977 International Microelectronics Symposium, p. 120–6.

565. McBride, D. G., "New interconnections for microelectronics", Des. Eng., 49, 51 (1976), EEA79–37742.

566. Markstein, H. W., "Fiber optics for electronic interconnec-
     tion", Electron. Packag. & Prod., 17 (4), 34-6 (1977),
     EEA81-6301.
567. Moran, K. P., Pascuzzo, A. L. and Yacavonis, R. A., "Spring-
     loaded module connectors for mounting an array of modules on
     circuit board", IBM Tech. Disclosure Bull., 20 (9), 3434-5
     (1978), EEA82-111.
568. Muccino, F. R., "Pin/socket interconnect - vital element in
     small, high density package", Insul./Circuits, 22 (6), 39-43
     (1976), EEA79-36697.
569. Ng, C. S. K., "Chip to pin carrier interconnection systems",
     IBM Tech. Disclosure Bull., 21 (7), 2707-8 (1978).
570. Nickols, S. E. and Patel, H. N., "Fabricating charge plate
     conductor and charge tunnel", IBM Tech. Disclosure Bull.,
     19 (12), 4752 (1977), EEA81-10041.
571. Patrick, R. and Schumacher, W., "Connector and cable consider-
     ations for high-speed digital circuits", 1975 WESCON Technical
     Papers, Paper 8/4, 8 pp., EEA79-16908.
572. Pavlik, J. and Thomas, D. L., "Increased density connector for
     the low end card on board package", IBM Tech. Disclosure Bull.,
     20 (9), 3401-2 (1978), EEA82-629.
573. Poole, J. R., "Connectors for electronic packaging", Electron.
     Prod. Methods & Equip., 6 (2), 11, 13 (1977), EEA80-41061.
574. Rose, A. S., Scheline, F. E. and Sikina, T. V., "Metallurgical
     considerations for beam tape assembly", 27th Electronic Com-
     ponents Conference, 1977, p. 130-4, EEA80-35961.
575. Ross, M. I., "High-density interconnections - an alternative
     to wire wrapping", Insul./Circuits, 23 (2), 33-5 (1977).
576. Savitsky, E. M., "Polyakova, V. P., Gorina, N. B., Dostanko,
     A. P., Chistyakov, J. D. and Zelenkov, V. A., "Thin film PdIn
     compound for integrated circuit metallization", J. Electron.
     Mater., 5 (3), 341-9 (1976), EEA79-46353.
577. Scaminaci, J., Jr., "Solderless press-fit interconnectors: A
     mechanical study of solid and compliant contacts", IEEE
     Trans. Manuf. Technol., MFT-6 (2), 23-30 (1977), EEA81-5595.
578. Seraphim, D. P., "Chip-module package interfaces", IEEE
     Trans. Components, Hybrids & Manuf. Technol., CHMT-1 (3),
     305-9 (1978), EEA81-49515.
579. Shoquist, M. C., "An advanced fiber-optic processor intercon-
     nection system", 14th IEEE Computer Society International Con-
     ference, COMPCON 77, Digest of Papers, p. 248-52.
580. Stephans, E., "Pinless module connector", IBM Tech. Disclosure
     Bull., 20 (10), 3872 (1978).
581. Turner, P., "Interconnection and packaging: no divorce possi-
     ble", Electron, no. 98, 35-6, 38, 41 (1976), EEA79-31895.
582. Van Der Drift, A. and Nickl Van Nikelsberg, K. F., "Hybrid
     interconnection of ICs by means of patterns on flexible tape",
     Electrocompon. Sci. & Technol., 4 (2), 53-6 (1977), EEA81-5629.

583. Widmann, D. W., "Metallization for integrated circuits using a lift-off technique", IEEE J. Solid-State Circuits, SC-11 (4), 466-71 (1976), EEA79-37683.
584. Wittmann, J. E., "Gold dot - a nonconnector approach for flat cable interconnections", Proc. Tech. Program Natl. Electron Packag. Prod. Conf., 1977, p. 483-91.
585. Yao, Y. L., "Data bus arrangement for Josephson tunneling device logic interconnections", Patent USA 4107554, Publ. August 1978.

## 2. Multilayer

586. Bergeron, R. J., "Double lift-off via interconnections and passivation process", IBM Tech. Disclosure Bull., 21 (4), 1371-2 (1978).
587. Bojc, F., "Multilayer interconnection techniques in electronic equipment", Elektroteh. Vestn., 44 (5), 302-4 (1977), Slovenian, EEA81-37129.
588. Breuninger, K., Haberland, D. R., Strasser, B. and Weitze, A., "Conductor crossovers provided by a combined thin-film and thick-film technology", Feinwerktech. & Messtech., 84 (7), 321-2 (1976), German, EEA80-6426.
589. Burns, J. A. and DiLeo, D. A., "Batch bonded crossovers for thin film circuits. I. Development", Solid State Technol., 19 (7), 26-31, 44 (1976), EEA79-37655.
590. Burr, R. P. and Hammond, J. P., "A unique interconnection technology for high speed logic", 1975 WESCON Technical Papers, Paper 8/3, 10 pp., EEA79-16907.
591. Carden, G. and McBride, D. G., "Wrap-around printed-circuit flat coil", IBM Tech. Disclosure Bull., 19 (12), 4547-8 (1977), EEA81-10007.
592. Corbin, V. R., Hitchner, J. E., Patnaik, B. and Ting, C. Y., "Forming self aligned via holes in thin film interconnection systems", Patent USA 4070501, Publ. January 1978.
593. Croset, M., "Flat connection for a semiconductor multilayer structure", Patent USA 3983284, Publ. September 1976, CA85-185710.
594. Currier, R. F., "Optimum procedures for fabricating rigid-flexible multilayer boards", Insul./Circuits, 24 (5), 29-31 (1978), EEA81-45725.
595. Ecker, M. E. and Olson, L. T., "Semiconductor package structure", IBM Tech. Disclosure Bull., 20 (8), 3092-3 (1978).
596. Engelmaier, W. and Kessler, T., "Investigation of agitation effects on electroplated copper in multilayer board plated-through holes in a forced-flow plating cell", J. Electrochem. Soc., 125 (1), 36-43 (1978), EEA81-19144.
597. Gardner, F. H. and Shaheen, J. M., "The dielectric film interconnect", Proceedings of the 1977 International Microelectronics Symposium, p. 93-9.

598. Gibbs, S. R. and Chow, K., "Multilevel metalization process", Patent USA 4045302, Publ. August 1977, CA87-176459.

599. Gniewek, J., Logan, J. S., Mauer, J. L. and Zielinski, L. B., "Dual insulators for planar multilevel interconnections", IBM Tech. Disclosure Bull., 21 (3), 1052-3 (1978).

600. Greer, S. E., "Fabrication of solid via structures in organic polymers", IBM Tech. Disclosure Bull., 19 (3), 911-12 (1976).

601. Gregoritsch, A. J., Jr., "Double-level metallurgy defect study", IEEE Trans. Electron Devices, ED-26 (1), 34-7 (1979).

602. Hampy, R. E., "Versatile $Ta_2$-N-W-Au/$SiO_2$/$Al$/$SiO_2$ thin film hybrid microcircuit metallization system", IEEE Trans. Parts, hybrids & Packag., PHP-11 (4), 263-72 (1975).

603. Hampy, R. E., Knauss, G. L., Komarek, E. E., Kramer, D. K. and Villanueva, J., "Processes and procedures for a thin film multilevel hybrid circuit metallization system based on W-Au/$SiO_2$/$Al$/$SiO_2$", Report SAND-76-0084, Sandia Labs., Albuquerque, N. Mex., 1976, 38 pp.

604. Hanazono, M., Asai, O. and Tamura, K., "Method of forming deposition films for use in multilayer metallization", Patent USA 4024041, Publ. May 1977.

605. Hibbs, T. J., "Multiwire techniques", Trans. Inst. Met. Finish, 55 (pt. 1), 13-16 (1977).

606. Hopfer, S., "Multi-layered MIC's", Microwave Syst. News, 5 (5), 67-8 (1975), EEA79-16190.

607. Johnson, A. H., "High conductivity ground planes for epoxy board", IBM Tech. Disclosure Bull., 21 (2), 558-9 (1978).

608. Kiessling, G., "New crossovers using a combined film technique", NTG-Fachber., no. 60, 76-81 (1977), German, EEA81-23621.

609. Kircher, C. J. and Zappe, H. H., "Wiring running above and below a ground plane existing on one side of a substrate", Patent German 2713532, Publ. January 1978, CA88-82757.

610. Kocsis, A., "A newly developed multilayer interconnection technique (thick film screening)", Electron. & Microelectron. Ind., no. 209, 62-3 (1975), EEA79-8980.

611. Kumar, A. H. and Miller, L. F., "Stress reduced conductive vias", IBM Tech. Disclosure Bull., 21 (1), 144 (1978).

612. Lennon, P. S. and Thomas, E. L., "Re-establishing interconnections in multilayer printed circuit structures", IBM Tech. Disclosure Bull., 21 (3), 980-1 (1978).

613. Martin, J. H., "Interconnection for planar electronic circuits", Report AD-D004761/3SL, Department of the Air Force, Washington, D. C., 1977, 12 pp.

614. Montier, M. and Boitel, M., "Interconnections by one or two layers for LSI-MOS using anodization process", International Conference on Microlithography, 1977, p. 349, French, EEA80-41591.

615. Morris, G. V., "Interconnecting layers in flexible circuits: solder interconnect alternative to P-T-H", Circuits Manuf., 17 (11), 49-50 (1977).

616. Mukai, K., Saiki, A., Harada, S. and Shoji, S., "High packing
     linear integrated circuits using planar metallization with
     polymer", 1977 International Electron Devices Meeting, Tech-
     nical Digest, p. 16-19, EEA81-24049.
617. Patterson, F. K., "Copper metallization and low K dielectric
     system for thick film multilayer interconnections", Insul./
     Circuits, 23 (13), 13-16 (1977), EEA81-19160.
618. Rau, B. R., "New philosophy for interconnection on multilayer
     boards", Report SU-326-P.39-13, Stanford Univ., Calif., 1977,
     7 pp.
619. Saiki, A., Harada, S., Okubo, T., Mukai, K. and Kimura, T.,
     "New transistor with two-level metal electrodes", J. Electro-
     chem. Soc., 124 (10), 1619-22 (1977).
620. Sarnacki, F. H., "Multilayer PC connector", IBM Tech. Dis-
     closure Bull., 20 (6), 2171-2 (1977), EEA82-627.
621. Schwartz, G. C. and Platter, V., "An anodic process for forming
     planar interconnection metallization for multilevel LSI", J.
     Electrochem. Soc., 122 (11), 1508-16 (1975), EEA79-9012.
622. Schwartz, G. C. and Platter, V., "Anodic processing for multi-
     level LSI", J. Electrochem. Soc., 123 (1), 34-7 (1976),
     EEA79-16885.
623. Schwartz, G. C. and Platter, V., "Monolithic studs as inter-
     level connectors in planar multilevel LSI", J. Electrochem.
     Soc., 123 (2), 300-1 (1976), EEA79-20539.
624. Skinner, D. W., "Wafer circuit package", Patent USA 3908155,
     Publ. September 1975.
625. Smith, M. C., "Multilevel electroforming", IBM Tech. Disclosure
     Bull., 20 (7), 2751-2 (1977), EEA81-45182.
626. Sproull, J. F., Gerry, D. J. and Bacher, R. J., "A high per-
     formance gold/dielectric/resistor multilayer system", Proceed-
     ings of the 1977 International Microelectronics Symposium,
     p. 20-4.
627. Sproull, J. F., Gerry, D. J. and Bacher, R. J., "High perfor-
     mance thick film system: gold/dielectric/resistor for multi-
     layers", Circuits Manuf., 17 (11), 14, 16, 18-20 (1977).
628. Stein, S. J., Spadafora, L. and Huang, C., "New developments
     in thick film conductors", NTG-Fachber., no. 60, 51-64
     (1977), German, EEA81-23638.
629. Stott, C., "Modular multiway connection system", New Electron.,
     11 (2), 54, 56 (1978), EEA81-28125.
630. Thomas, J. H., III, Morabito, J. M. and Lesh, N. G., "Ti-Cu-
     Ni-Au (TCNA) compatibility with resistor and bi-level cross-
     over circuit processing", J. Vac. Sci. & Technol., 13 (1),
     152-5 (1976), EEA79-24879.
631. Traskos, R. T., "A flex circuit composite based on a fiber
     blend", Proceedings of the 13th Electrical/Electronics Insu-
     lation Conference, 1977, p. 72-4, EEA81-14290.
632. Voida, G., "Solder fused interconnections in multilayer cir-
     cuits", Insul./Circuits, 23 (10), 77-82 (1977), EEA81-899.

633. Zielinski, L. B., "Passivated metal interconnection system with a planar surface", Patent USA 3985597, Publ. October 1976, CA85-185717.

VI.  METALLIZED CERAMICS

634. Anon., "Can copper ceramic metalization substitute for thick film paste?", Circuits Manuf., 16 (10), 40, 42, 44 (1976), EEA80-3291.
635. Antonucci, R. F. and Lahey, E. L., "Removal of potassium permanganate stain", IBM Tech. Disclosure Bull., 21 (7), 2909 (1978).
636. Belikov, V. N., Belousenko, A. P. and Vyskrebtsev, V. P., "Metalized ceramic articles", Patent USSR 590306, Publ. January 1978, CA88-125381.
637. DeVore, J. A., "Metallization of beryllia ceramics for microelectronic use", Proceedings of the 13th Electrical/ Electronics Insulation Conference, 1977, p. 70-1, EEA81-14313.
638. Funari, J., Myers, F. R. and Thomson, W. G., "Flanged pin metallized ceramic substrate design", IBM Tech. Disclosure Bull., 21 (1), 94-5 (1978).
639. Gedney, R. W., Gow, J., III and Rasile, J., "MC substrate process for optimum chip-joining yields", IBM Tech. Disclosure Bull., 20 (9), 3395 (1978), EEA82-650.
640. Gedney, R. W. and Rasile, J., "Integrated circuit package", Patent USA 4072816, Publ. February 1978.
641. Gedney, R. W. and Rodite, R. R., "Metallized ceramic and printed circuit module", Patent USA 4082394, Publ. April 1978.
642. Gedney, R. W. and Webizky, G. G., "Low-cost integrated circuit", IBM Tech. Disclosure Bull., 20 (9), 3399-400 (1978).
643. Goss, B. R., "Metallization of beryllia composites", Patent USA 3993821, Publ. November 1976.
644. Gow, J., Hoffman, H. S., Hughes, J. J. and Thomas, G. M., "Modified chromium dam for metallized ceramic metallurgy", IBM Tech. Disclosure Bull., 20 (10), 3856 (1978).
645. Gow, J. and Stephans, E., "Multilayer ceramic expander", IBM Tech. Disclosure Bull., 21 (2), 498-9 (1978).
646. Harris, D. H., "Are plasma sprayed metallization on microwave ceramic substrates", 12th Electrical/Electronic Insulation Conference, 1975, p. 99-101, EEA79-46553.
647. Hawrylo, F. Z. and Kressel, H., "Metalized device", Patent USA 3945902, Publ. March 1976.
648. Hoffman, H. S., "Post evaporation annealing of chromium-copper-chromium metallized ceramics", Proceedings of the Electronic Components Conference, Vol. 26, 1976, p. 173-6, CA85-147430.
649. Holcombe, C. E. and Snyder, W. B., "Gold layers on ceramics", J. Less-Common Met., 51 (1), 163-4 (1977), EEA80-12853.

650. Mahapatra, S. and Prasad, S. N., "A new electroless method for low-loss microwave integrated circuits", IEEE Trans. Components, Hybrids & Manuf. Technol., CHMT-1 (4), 428-31 (1978).

651. Mase, S. and Watanabe, T., "Multilayer metalized beryllia ceramics", Patent USA 3927815, Publ. December 1975.

652. Mentone, P. F., "Selective metalization of ceramic nonconductors", Patent UK 1429706, Publ. March 1976.

653. Pincus, A. G. and Chang, S. H., "Reminiscences about metallized ceramics and their use in seals", Am. Ceram. Soc. Bull., 56 (4), 433-36 (1977), CA87-10203.

654. Pottier, B., "Pretinning the metallurgy of a module to enhance wettability", IBM Tech. Disclosure Bull., 20 (2), 622-3 (1977), EEA81-13295.

655. Tutorskaya, N. N., Yushkina, E. T., Pletneva, L. A., Friborg, V. E. and Chebotarev, A. Y., "Solder for soldering metallized ceramic to metal", Patent USSR 564128, Publ. July 1977, CA87-121679.

656. Van Vestrout, V. D., "Hybrid module including thin-film conductors and paste resistors", IBM Tech. Disclosure Bull., 19 (12), 4541 (1977), EEA81-10050.

657. Van Vestrout, V. D., "Power module with high power dissipation", IBM Tech. Disclosure Bull., 19 (9), 3346-7 (1977), EEA80-35001.

VII.   ENCAPSULATION

658. Anon., "Plastics encapsulated integrated circuits. Pros and cons as seen by the manufacturer", Elektron. J., 11 (12), 28-30 (1976), German, EEA80-27987.

659. Anon., "Low cost encapsulation process", Polym. News, 3 (2), 91-2 (1976), EEA80-31397.

660. Anon., "Using sealants, encapsulants and adhesives", Electron. Prod., 6 (4), 37, 39 (1977), EEA81-9246.

661. Baker, T. E., Fix, G. L. and Judge, J. S., "The application of polyurethane encapsulants for electronic packages", Electrochemical Society Spring Meeting, Extended Abstracts, 1977, p. 125-7, EEA81-4781.

662. Baker, T. E. and Judge, J. S., "Considerations in the selection of materials for encapsulating high-speed computer assemblies", Electrochemical Society Spring Meeting, Extended Abstracts, 1975, p. 302-4, EEA79-91.

663. Brady, M. J., "Laser diode facet encapsulation and protection", IBM Tech. Disclosure Bull., 21 (4), 1747 (1978).

664. Dargavel, G., "Packaging electronic circuits. I. Casting and coating", Electron, no. 103, 49-50 (1976), EEA80-40.

665. Donaldson, P. E. K., "The encapsulation of microelectronic devices for long-term surgical implantation", IEEE Trans. Biomed. Eng., BME-23 (4), 281-5 (1976), EEA79-30534.

666. Dumaine, G., Durand, J., Lemoine, J. M. and Pottier, B., "Deposit of a preform for encapsulating multilayer ceramic modules", IBM Tech. Disclosure Bull., 20 (4), 1470 (1977), EEA81-23647.

667. Fairchild Camera and Instrument Corp., "Packaging and epoxy resins for packaging electrical and electronic components", Patent Japan 75-144758, Publ. November 1975, CA85-136065.

668. Fox, M. J., "A comparison of the performance of plastic and ceramic encapsulations based on evaluation of CMOS integrated circuits", Microelectron. & Reliab., 16 (3), 251-4 (1977), EEA81-136.

669. Gibney, J. T., Jr., Jaeger, T. H. and Wirtz, L. H., "Attachment of protective caps", IBM Tech. Disclosure Bull., 19 (7), 2646-7 (1976), EEA80-32252.

670. Gnedovets, A. G., Zakirov, R. G. and Zuev, I. V., "Calculation of the temperature of the shell of integrated circuits in encapsulation by roll seam welding", Weld. Prod., 24 (4), 32-4 (1977), EEA81-32235.

671. Gontscharenko, J. W. and Gesemann, R., "Protection against humidity for electronic components by means of plastic materials", Hermsdorfer Tech. Mitt., 17 (47), 1500-3 (1977), German, EEA80-27988.

672. Greer, S. E., "Organic cap for first level packages", IBM Tech. Disclosure Bull., 20 (2), 583-4 (1977), EEA81-18367.

673. Hakim, E. B., "A case history: procurement of quality plastic encapsulated semiconductors", Solid State Technol., 20 (9), 71-3 (1977), EEA81-14685.

674. Hartley, P., "Encapsulation systems stabilise electronics", Des. Eng., 61 (1977), EEA81-132.

675. Jaffe, D. and Soos, N. A., "Encapsulation of integrated circuits", Patent USA 4017495, Publ. April 1977, CA86-198873.

676. Jumbeck, G. I., "Removable multichip module cap", IBM Tech. Disclosure Bull., 20 (1), 73-4 (1977), EEA81-14354.

677. Lawson, R. W. and Harrison, J. C., "Plastic encapsulation of semiconductors", Plastics in Telecommunication, 1st International Conference, Paper 6, 30 pp.

678. Luettge, H. J., "Encapsulated active devices for non-hermetic thick/thin film hybrids", International Electrical, Electronics Conference and Exposition, 1977, p. 134, EEA81-23645.

679. Magdo, S., "Semiconductor encapsulation", IBM Tech. Disclosure Bull., 20 (10), 3903-5 (1978).

680. Manfre, E., "Welding & sealing microcircuit packages", Circuits Manuf., 16 (4), 32, 34, 36 (1976), EEA79-37613.

681. Marchetti, R. D., "Automatic liquid resin molding is a 'now' in thermosets", SPE, Chicago Section, Regional Technical Conference, Proceedings, 1976, p. 109-15.

682. Nenyei, Z., "Casing of semiconductor devices. I.", Finommech. & Mikrotech., 15 (10), 289-98 (1976), Hungarian, EEA80-12539.

683. Pettican, K. A., "Plastics encapsulation of electronic compo-
     nents", Electrotechnology, 4 (2), 12-15 (1976), EEA79-19573.
684. Price, S. J., Johnson, R. L. and Chapman, J. F., "For reliable,
     severe environment performance, encapsulate LED's with a clear
     silicone", Insul./Circuits, 23 (11), 53-6 (1977).
685. Richards, B. G., "Component encapsulation-fact and fiction",
     Electron. Prod. Methods & Equip., 4 (7), 64, 67-8 (1975),
     EEA79-3729.
686. Rosler, R. K., "Thermoset plastics for semiconductor encapsu-
     lation", Electron. Packag. & Prod., 18 (7), 168-7 (1978).
687. Sunda, J. A. and Taylor, J. R., "High performance in minimum
     volume (plastic encapsulated transistors)", Electron, no. 81,
     50, 53 (1975), EEA79-4676.

VIII.   HERMETICITY

688. Anon., "A three metal process for hermetically sealing ICs in
     plastic casings", Elektron. & Elektrotech., 30 (763-4), 230
     (1975), Dutch, EEA79-1044.
689. Avery, R. P. and Hutter, H. G., "Improved one piece SMA her-
     metically sealed launcher", 12th Electrical/Electronics Insu-
     lation Conference, 1975, p. 106-7, EEA79-46555.
690. Dietz, R. L., "New gold alloy for aluminium bonding in MOS
     hermetic packages", 27th Electronic Components Conference,
     1977, p. 367-73, EEA80-35969.
691. Doyle, K. B., "Some experiences and conclusions using soldered
     and welded packages for hermetic thick film hybrids", Micro-
     electron. & Reliab., 16 (4), 303-7 (1977).
692. Francis, J. F. and Orosz, E. M., "Forming a hermetic bond in
     semiconductor packages", IBM Tech. Disclosure Bull., 20 (8),
     3085-6 (1978).
693. Galakhov, N. V., Gevorkyan, G. K., Piliposyan, P. M. and
     Radchenko, A. V., "Investigating the hermetic sealing of
     strip-wound cores by the deposition method", Elektrotekhnika,
     no. 12, 57-8 (1975), Russian, EEA79-11898.
694. Gallace, L. J. and Pujol, H. L., "Hermeticity of COSMOS inte-
     grated circuits", New Electron., 10 (12), 83 (1977),
     EEA81-10407.
695. Greer, S. E., "Method of sealing polyamide organic laminates",
     IBM Tech. Disclosure Bull., 20 (5), 1749 (1977).
696. Gregor, L. V., Hirsch, H., Überbacher, G. and Walsh, T. J.,
     "Hermetic sealing of semiconductor packages", IBM Tech. Dis-
     closure Bull., 21 (7), 2821 (1978).
697. Hannan, J. J., Prasad, C., Shapiro, M. R. and Walker, W. J.,
     "Removable hermetic cap", IBM Tech. Disclosure Bull., 21 (2),
     574-5 (1978).
698. Howard, R. T., "Edge seal for multilevel integrated circuit
     with organic interlevel dielectric", IBM Tech. Disclosure
     Bull., 20 (8), 3002-3 (1978).

699. Kale, V. S., "Interaction of parylene and moisture in her-
     metically sealed hybrids", Proceedings of the 28th Electronic
     Components Conference, 1978, p. 344-9, EEA81-42162.
700. Kehagioglou, T., "Semiconductor module seal", IBM Tech. Dis-
     closure Bull., 20 (11), 4863 (1978).
701. Kim, C. K., "Method of hermetically sealing semiconductor
     devices", Report AD-D002559/3SL, Department of the Air Force,
     Washington, D. C., 1976, 8 pp.
702. Lamson, M. A. and Ramsey, T. H., Jr., "Improving the quality
     of glass-sealed ceramic dips", Electron. Packag. & Prod., 18
     (1), 84-96 (1978).
703. Miley, J. E. and Simpson, G., "Try impact extrusion for low-
     cost MIC packaging", Microwaves, 15 (6), 62, 66, 68 (1976),
     EEA79-36685.
704. Noren, S., "Recent developments in low cost hermetic packag-
     ing", International Microelectronics Conference, Proceedings
     of the Technical Program, 1975, p. 122-4.
705. Olesen, S. T., "Hermetic equipment casings", EC-Nyt, no. 48,
     8-9 (1976), Danish, EEA79-24046.
706. Pinnow, D. A., Wysocki, J. A. and Robertson, G. D., "Hermeti-
     cally sealed high strength fiber optical waveguides", Trans.
     Inst. Electron. Commun. Eng. Jpn. Sect. E, E61 (3), 171-3 (1978).
707. Rich, T. C. and Camlibel, I., "Simple optical fiber-to-metal
     hermetic seal", Am. Ceram. Soc. Bull., 57 (2), 234-5 (1978).
708. Rifkin, A. A., Schiller, J. M. and Turetzky, M. N., "Sealant
     for semiconductor package", IBM Tech. Disclosure Bull., 20
     (9), 3446 (1978).
709. Ruthberg, S., "Gas infusion into doubled hermetic enclosures",
     IEEE Trans. Parts, Hybrids & Packag., PHP-13 (2), 110-16
     (1977), EEA80-31395.
710. Tesny, W., "Air tight sealing in electronic devices", Elek-
     tronika, 19 (2), 63-4 (1978), Polish, EEA81-37131.
711. Villien, P., "Microwave integrated circuit process investiga-
     tion, work package-2, hermetic sealing", Report N77-18354/9SL,
     Danish Research Center for Applied Electronics, Hoersholm,
     1976, 14 pp.
712. Wilbur, C. V., "Hermetic seal for display cells", IBM Tech.
     Disclosure Bull., 20 (6), 2402 (1977).
713. Yasumura, G. and Krumme, J. F., "Hermetic package sealing with
     nitinol", International Microelectronics Conference, Proceed-
     ings of the Technical Program, 1975, p. 230-7.

IX.  RESISTORS

714. Ambrozy, A., "Variance noise", Electron. Lett., 13 (5), 137-8
     (1977), EEA80-13196.
715. Ameen, J. G. and Hoffman, H. S., "Trimming cermet resistors",
     IBM Tech. Disclosure Bull., 19 (7), 2485 (1976), EEA80-24611.

716. Anantha, N. G., Bhatia, H. S., Eardley, D. B. and Pogge, H. B.,
     "Epitaxial resistor with higher sheet resistance", IBM Tech.
     Disclosure Bull., 21 (7), 2805 (1978).
717. Anon., "Circuit elements. III. Fundamentals for practical ap-
     plications (resistors)", Funk-Tech., 32 (6), 59-63 (1977),
     German, EEA80-35779.
718. Anon., "Pots, trimmers, and resistors [EE's 1977/78 guide]",
     Eval. Eng., 16 (5), 10-12, 14-16 (1977), EEA81-19073.
719. Anon., "Thick film resistors with Polypyron", Radio Mentor
     Electron., 44 (3), 94 (1978), German, EEA81-37818.
720. Ansell, M. P., "Conduction processes in thick film resistors.
     I.", Electrocompon. Sci. & Technol., 3 (3), 131-40 (1976),
     EEA80-17038.
721. Ansell, M. P., "Conduction processes in thick film resistors.
     II.", Electrocompon. Sci. & Technol., 3 (3), 141-51 (1976),
     EEA80-17039.
722. Antes, A. C., Drehle, J. R. and Harrison, B. H., "Fabrication
     of thick film resistors", Patent USA 3998980, Publ. December
     1976.
723. Aritomo, A., Shirakawa, T., Yoshida, S. and Murasumi, T.,
     "Improvement of uniformity of sputtered film sheet resistance",
     27th Electronic Components Conference, 1977, p. 61-7,
     EEA80-15908.
724. Barlic, P., "Ni-Cr thin film resistors", Elektroteh. Vestn.,
     44 (3), 172-7 (1977), Slovene, EEA81-14314.
725. Barnwell, P. G., "Extending the use of thick film", Electron,
     no. 103, 56, 58 (1976), EEA80-494.
726. Batev, K. P. and Vacev, K. D., "Thin-film hybrid integrated
     circuits with higher accuracy of the ratio of integrated re-
     sistors", Elektro Prom.-st. & Priborostr., 11 (3), 101-3 (1976),
     Bulgarian, EEA79-33092.
727. Beensh-Marchwicka, G., Dembicka-Jellonkowa, S. and Krol-
     Stepniewska, L., "Aging mechanisms in thin film Cr, CrAu and
     $TiN_x$ resistors", Thin Solid Films, 36 (2), 361-3 (1976),
     PA79-83835.
728. Bergeron, D. L. and Stephens, G. B., "Measuring the voltage
     coefficient of resistance on integrated circuit resistors",
     IBM Tech. Disclosure Bull., 19 (9), 3495-7 (1977), EEA80-36400.
729. Bergeron, D. L. and Stephens, G. B., "Method of improving
     match on integrated resistors", IBM Tech. Disclosure Bull.,
     20 (3), 1089 (1977), EEA81-24034.
730. Berndlmaier, E., "Selectable resistance values for integrated-
     circuit resistors", IBM Tech. Disclosure Bull., 19 (12),
     4620-1 (1977), EEA81-10372.
731. Bibeau, W. E., Porter, W. A. and Parker, D. L., "Correlation
     of thin film adhesion with current noise measurements of
     $Ta_2N$-Cr-Au resistors on sapphire and alumina substrates",
     Proceedings of the 28th Electronic Components Conference,
     1978, p. 427-32, EEA81-42155.

732. Bonchar, N. R., Lysova, E. V., Popov, V. I. and Popova, A. A.,
     "Electrophysical parameters of vacuum-deposited resistors
     using copper-manganese-titanium alloys", Izv. VUZ Radioelek-
     tron., 20 (1), 76-8 (1977), Russian, EEA80-32241.
733. Boonstra, A. H. and Mutsaers, C. A. H. A., "Small values of
     the temperature coefficient of resistance in lead rhodate
     thick films ascribed on a compensation mechanism", Thin
     Solid Films, 51 (3), 287-96 (1978), EEA81-45762.
734. Bos, L. W., Ostrander, W. J. and Ver Hoef, J., "A new center-
     tapped medium-power microfilm resistor", IEEE Trans. Parts,
     Hybrids & Packag., PHP-12 (13), 212-17 (1976), EEA80-419.
735. Bruck, D. B. and Pollens, A. L., "Why the design nod goes to
     resistors made as thin-film monolithic networks", Electronics,
     51 (16), 99-104 (1978), EEA81-42070.
736. Buczek, D. M., "Thin-film NiCr resistor", J. Vac. Sci. &
     Technol., 15 (2), 370-2 (1978), EEA81-42074.
737. Buzan, F. E., Grier, J. D., Sr., Bertsch, B. E. and Thoryk, H.,
     "A thick film base metal resistor and compatible hybrid sys-
     tem", Electrocompon. Sci. & Technol., 5 (1), 15-25 (1978),
     EEA81-37909.
738. Carcia, P. F., "Resistors and compositions therefor", Patent
     USA 3974107, Publ. August 1976.
739. Carcia, P. F., Champ, S. E. and Flippen, R. B., "High voltage
     stable thick film resistors", 26th Electronic Components Con-
     ference, 1976, p. 156-62, EEA79-37649.
740. Carcia, P. F. and Rosenberg, R. M., "Thick film resistor series
     features low contact noise, high voltage stability, and low
     process sensitivity", Insul./Circuits, 22 (10), 25-7 (1976),
     EEA80-498.
741. Carter, J. M., "Metal film resistors, the state of the art",
     Electron. Equip. News, 10-12 (1977), EEA81-28053.
742. Cattaneo, A., Cocito, M., Forlani, F. and Prudenziati, M.,
     "Influence of the metal migration from screen-and-fixed ter-
     minations on the electrical characteristics of thick-film
     resistors", Electrocompon. Sci. & Technol., 4 (3-4), 205-11
     (1977), EEA81-19158.
743. Chen, T. M. and Rhee, J. G., "The effects of trimming on the
     current noise of thick film resistors", Solid State Technol.,
     20 (6), 49-53 (1977), EEA80-35917.
744. Coleman, M. V., "The effects of $SO_2$ and $H_2S$ atmospheres on
     thick film resistors", Microelectron. & Reliab., 14 (5-6),
     445-6 (1975), EEA79-24896.
745. Collins, F. M., "Low-cost precision resistor networks", 26th
     Electronic Components Conference, 1976, p. 313-16, EEA79-37659.
746. Counts, W. E., "Resistor stability and power dissipation",
     Solid State Technol., 20 (10), 59-62 (1977), EEA81-19162.
747. Croson, E. B., "Chromium-silicon oxide thin film resistors",
     Patent USA 3996551, Publ. December 1976.

748. De Sabata, I., "Reciprocity theorems for a nonstationary elec-
     trokinetic field and the RC parameters of the resistors and
     capacitors", Bul. Stiint. & Teh. Inst. Politeh. 'Traian Vuia'
     Timisoara, 21 (1), 59-62 (1976), French, EEA80-28547.
749. Dearden, J., "Electroless plating - its applications in
     resistor technology", Electrocompon. Sci. & Technol., 3 (2),
     103-11 (1976), EEA80-417.
750. Dearden, J., "Advances in high voltage resistor design",
     Electron, no. 132, 26, 30 (1978), EEA81-28055.
751. Della Mea, G., Baeri, P., Campisano, S. U. and Cocito, M.,
     "Determination of oxygen and nitrogen in tantalum thin films
     by nuclear techniques and Auger spectroscopy", Thin Solid
     Films, 51 (1), L5-8 (1978), EEA81-42154.
752. Domingos, H. and Wunsch, D. C., "High pulse power failure of
     discrete resistors", IEEE Trans. Parts, Hybrids & Packag.,
     PHP-11 (3), 225-9 (1975).
753. Dubowitzkaya, I. M. and Kasarin, I. G., "Reproducibility of
     thin film resistors", Nachrichtentech. Elektron., 27 (11),
     469-70 (1977), German, EEA81-14172.
754. Dutta Roy, S. C. and Das, V. G., "On resistor loops in hybrid
     circuits", Proc. IEEE, 65 (4), 583 (1977).
755. Elmgren, J. A. and Memis, I., "Thick-film resistors", IBM
     Tech. Disclosure Bull., 19 (7), 2484 (1976), EEA80-24616.
756. Engel, S., "Selection of resistors", Elektronik, no. 3, 14,
     16 (1977), Danish, EEA80-35774.
757. Faith, T. J., Jr., "High power density thin film resistors",
     IEEE Trans. Parts, Hybrids & Packag., PHP-11 (4), 273-81
     (1975), EEA79-13007.
758. Faith, T. J., Jr. and Jennings, J. W., "Fluid abrasive trim-
     ming of thin-film resistors", IEEE Trans. Parts, Hybrids &
     Packag., PHP-12 (2), 133-8 (1976), EEA79-37653.
759. Ferraris, G. P., "Thin film resistors for high precision net-
     works", Thin Solid Films, 38 (1), 21-33 (1976), EEA80-20538.
760. Forlani, F. and Prudenziati, M., "Electrical conduction by
     precolation in thick film resistors", Electrocompon. Sci. &
     Technol., 3 (2), 77-83 (1976), EEA80-495.
761. Forster, J., "Reactive formation of resistive coatings", Radio
     Mentor Electron., 42 (9), 342-6 (1976), German, EEA80-422.
762. Fu, S. L., Liang, J. M., Shiramatsu, T. and Wu, T. S., "Effects
     of glass modifiers on the electrical properties of thallium
     oxide thick film resistors", Mater. Res. Bull., 12 (6), 569-76
     (1977), EEA80-35915.
763. Goswami, A. P., Satyanarayana, L. and Sivaji, N. V. M., "A
     screening technique for establishing the stability of metal
     film resistors", Electrocompon. Sci. & Technol., 5 (2),
     71-7 (1978), EEA81-42071.
764. Gotra, Z. Y., Balahin, M. Y. and Kuzkin, A. P., "Tungsten
     thin film resistors", Prib. & Sist. Upr., no. 12, 48-9 (1976),
     Russian, EEA80-20579.

765. Gotra, Z. Y. and Kuzkin, A. P., "Rhenium thin-film resistors for analogue measurements", Prib. & Sist. Upr., no. 11, 47-8 (1975), Russian, EEA79-20515.

766. Grierson, M. and Chen, T. M., "Application of computerized current flux plots in resistor noise calculations", Proceedings of the Southeastcon '78 Region 3 Conference, p. 175-8, EEA81-42157.

767. Griessing, J., "Reactive sputtering of NiCr resistors with closely adjustable temperature coefficient of resistance", Electrocompon. Sci. & Technol., 4 (3-4), 133-7 (1977), EEA81-19153.

768. Groth, L., "Thin film resistor networks in hybrid circuits", Solid State Technol., 20 (3), 45-9 (1977), EEA80-24626.

769. Gusev, Y. A., Oleinik, A. V. and Anikin, A. Y., "Film strain resistor for temperatures up to 1000°C", Instrum. & Exp. Tech., 19 (2), 564-6 (1976), EEA80-20532.

770. Guttensohn, A. E., "A realistic approach to thick film resistor design", Solid State Technol., 21 (9), 73-8 (1978), EEA82-646.

771. Hajdu, L., "The effect of the trimming technique for top hat type resistors on the design process", Finommech. & Mikrotech., 14 (9), 275-7 (1975), Hungarian, EEA79-8985.

772. Haribanova, V., "On some technological aspects of thick resistor films on the basis of Pd, Ag and glass", Elektrotech. Cas., 29 (3), 205-8 (1978), Slovak, EEA81-33022.

773. Harman, R., Kempny, M., Vanet, O. and Tvarozek, V., "Cathode-sputtered resistor layers for hybrid integrated circuits", Nachrichtentech. Elektron., 26 (10), 395-7 (1976), German, EEA80-3306.

774. Hart, P. J., "Resistor networks in hybrid circuits", Engineering, 217 (6), 492-4 (1977).

775. Hart, P. J., "Resistor stability and power dissipation", Electron, no. 132, 21-2, 25 (1978), EEA81-28180.

776. Haynes, K. L. and Ruwe, V. W., "Experimental characterization of thick-film resistor stability and performance", Proceedings of the 1977 International Microelectronics Symposium, p. 76-80.

777. Hazell, J., "The growth in demand for thick film resistor networks in standard configurations", Electron, no. 114, 43, 45 (1977), EEA80-39514.

778. Headley, R. C., "Designing for minimum resistor current noise", Electron. Packag. & Prod., 17 (12), 111-12, 114, 116 (1977), EEA81-33023.

779. Hellstrom, S. and Wesemeyer, H., "Nonlinearity measurements of thin films", Vacuum, 27 (4), 339-43 (1976), EEA81-5617.

780. Heribanova, V., "Technological aspects of thick film resistor on the basis of palladium, silver and glass", Elektrotech. Cas., 29 (1), 83-7 (1978), EEA81-28179.

781. Hinuber, W. and Fendler, R., "Foil techniques for the production of multichip hybrid wiring carriers with integrated resistors", Nachrichtentech. Elektron., 28 (4), 149-50 (1978), German, EEA81-33029.

782. Hirakawa, H., Oide, K. and Ogawa, Y., "Electronic cooling of resistors", J. Phys. Soc. Jpn., 44 (1), 337-8 (1978), EEA81-14171.

783. Hoffman, H. S. and Stephans, E., "Thin film resistor design", IBM Tech. Disclosure Bull., 21 (6), 2279-80 (1978).

784. Hollendus, J., "Designing of thin-film resistors and automatic value adjustment by the help of a computer", Finommech. & Mikrotech., 15 (10), 308-11 (1976), Hungarian, EEA80-17036.

785. Howell, G. and Winebarger, J., "Don't overlook self-heating of resistors", Electronics, 50 (17), 117-9 (1977), EEA80-40992.

786. Hsieh, K. C., Chenette, E. R. and Van Der Ziel, A., "Current noise in surface layer integrated resistors", Solid State Electron., 19 (6), 451-3 (1976).

787. Huang, C. Y. D., Gelb, A. S. and Stein, S. J., "Thick-film resistor characteristics under severe voltage stress", Proceedings of the 1977 International Microelectronics Symposium, p. 65-71.

788. Illyefalvi-Vitez, Z., "Trimming of thin film resistors", Finommech. & Mikrotech., 14 (9), 265-8 (1975), Hungarian, EEA79-8990.

789. Illyefalvi-Vitez, Z., "Value adjustment of thin-film resistors by changing planar forms", Finommech. & Mikrotech., 15 (3), 69-75 (1976), Hungarian, EEA79-33094.

790. Johnston, P. E., "The performance and application of thick-film chip resistors on PC boards", 27th Electronic Components Conference, 1977, p. 110-15, EEA80-35783.

791. Jones, W., "Zero temperature coefficient of resistance bi-film resistor", Report AD-D003810/9SL, Department of the Navy, Washington, D. C., 1977, 11 pp.

792. Kasabov, I. D., Petrov, I. N., Nencheva, G. P. and Georgiev, V. K., "High value junction microresistors prepared by double ion implantation of $^{11}B$ and of noble gas ions", Bulg. J. Phys., 3 (2), 166-73 (1976), EEA80-9126.

793. Kawamata, E. and Kitamura, Y., "Transparent conducting film resistor produced by thermal decomposition of tetraalkyltin", Oyo Buturi, 44 (5), 474-82 (1975), Japanese, EEA79-8994.

794. Keenan, W. F., "Pulsed overload tolerance of Si/Cr, Ni/Cr and Mo/Si thin film resistors on integrated circuits", IEEE Trans. Reliab., R-25 (4), 248-53 (1976).

795. Kirby, P. L., "Applications of resistive thin films in electronics", Thin Solid Films, 50, 211-21 (1978), EEA81-42153.

796. Kolonits, V. P., Strausz, T. and Koltai, M., "New reaction kinetic aspects of the thermal oxidation of sputtered tantalum (nitride) layers", Vacuum, 27 (12), 635-42 (1977), EEA81-23525.

797. Kusy, A., "On the structure and conduction mechanism of thick
     resistive films", Thin Solid Films, 37 (3), 281-302 (1976),
     EEA80-6430.
798. Kusy, A., "Chains of conducting particles that determine the
     resistivity of thick resistive films", Thin Solid Films, 43
     (3), 243-50 (1977), EEA80-35918.
799. Kusy, A., "1/f noise of ruthenate films", Arch. Elektrotech.,
     27 (1), 49-62 (1978), Polish, EEA81-45759.
800. Lahiri, S. K., "A thin film resistor for Josephson tunneling
     circuits", Thin Solid Films, 41 (2), 209-15 (1977), EEA80-24538.
801. Langley, R. A., "Studies of tantalum nitride thin film resis-
     tors", Report SAND-74-7006, Sandia Labs., Albuquerque, N.
     Mex., 1975, 23 pp.
802. Larry, J. R., Popowich, M. J., Headley, R. C. and Rosenberg,
     R. M., "A new high stability resistor system", 27th Electronic
     Components Conference, 1977, p. 348-57, EEA80-35784.
803. Law, J. T., "The role of thin film hybrid microcircuits in
     electronics", Thin Solid Films, 36 (2), 323-9 (1976),
     EEA79-41606.
804. Lockert, C., "Thick film resistors for kilovolt circuits",
     Mach. Des., 49 (28), 156-60 (1977), EEA81-45673.
805. Lott, J., "Resistors [on Canadian market]", Can. Electron.
     Eng., 21 (11), 36 (1977), EEA81-23514.
806. Loudon, R., "Thick-film resistor networks reduce digital system
     costs", Electron, no. 84, 21-2, 25 (1975), EEA79-8982.
807. Luby, S. and Chromik, S., "Trimming of $Ta_xN$ thin film resis-
     tors by anodic oxidation", Elektrotech. Cas., 28 (2), 140-3
     (1977), Slovak, EEA80-24609.
808. McCormick, W. and Tan, S. I., "Vitreous, high temperature,
     film resistor material", IBM Tech. Disclosure Bull., 20 (2),
     785 (1977), EEA81-19077.
809. Maddison, E. A., "Thick film resistor networks", Electron,
     no. 125, 35 (1977), EEA81-19156.
810. Maddison, E., "Thick film resistor technology", New Electron.,
     11 (5), 125-6 (1978), EEA81-28181.
811. Maloberti, F., Montecchi, F. and Svelto, V., "Flicker noise
     in thick film resistors", Alta Freq., 44 (11), 681-3 (1975).
812. Manolescu, A. M., "Aspects regarding the design of thin film
     resistances", Electrotech. Electron. & Autom. Autom. &
     Electron., 20 (2), 57-64 (1976), Rumanian, EEA80-418.
813. Martin, F. W., "Low firing polymer thick film enables the
     screen printing of resistors and conductors on PC boards",
     Insul./Circuits, 21 (8), 29-32 (1975).
814. May, E. J. P. and Sellars, W. D., "1/f noise produced by
     radio-frequency current in resistors", Electron. Lett., 11
     (22), 544-5 (1975).
815. Mayer, G., "Integration of nickel-chromium resistors and
     silicon oxide capacitors", Funk-Tech., 31 (19), 604-14
     (1976), German, EEA80-9194.

816. Memis, I. and Rasile, J., "Fabrication of cermet resistors", IBM Tech. Disclosure Bull., 20 (3), 957-8 (1977), EEA81-19078.
817. Moran, P. L. and Maiti, C. K., "A program to predict the resistance of trimmed film resistors", Electrocompon. Sci. & Technol., 3 (3), 153-64 (1976), EEA80-17035.
818. Mueller, O., "Comment on 1/f noise in resistors", Electron. Lett., 12 (2), 48-9 (1976).
819. Mulligan, W., "Need a high-voltage resistor?", Electron. Des., 25 (20), 84-7 (1977), EEA81-23516.
820. Naguib, H. M., "The characterization of thick-film resistors terminated with Pd/Ag conductors of various compositions", Proceedings of the 1977 International Microelectronic Symposium, p. 48-59.
821. Nakada, Y. and Schochk, T. L., "Reliability study of a high-precision thick film resistor network", IEEE Trans. Parts, Hybrids & Packag., PHP-13 (3), 229-34 (1977), EEA81-5624.
822. Nakazawa, S., Shinada, T. and Shiomi, H., "Life-time acceleration for tantalum resistor under the humid environment", Bull. Electrotech. Lab., 41 (7), 556-61 (1977), Japanese, EEA81-5528.
823. Nielsen, G., "Will custom resistor networks fit into your next design?", EDN, 22 (5), 104-7 (1977), EEA80-24629.
824. Nordstrom, T. V., "Thermal degradation of thick film resistors", Electrochemical Society Spring Meeting, Extended Abstracts, 1977, p. 42-3, EEA81-5626.
825. Olumekor, L. and Beynon, J., "Reliability of pure Mn and Mn/MgF$_2$ cermet thin film resistors", Thin Solid Films, 44 (3), L17-18 (1977), EEA80-40996.
826. Ormond, T., "Variable resistors and networks – a look at what products are getting 'on board' today", EDN, 22 (4), 64-72, 74 (1977), EEA80-20529.
827. Ostwald, R. and Bogenschutz, A. F., "Compatibility of thick film resistors on glass ceramic layers", Ber. Dtsch. Keram. Ges., 53 (2), 55-8 (1976), German, EEA79-24894.
828. Packowski, E., "Forced testing methods for the evaluation of resistors' reliability", Elektronika, 18 (10), 405-8 (1977), Polish, EEA81-19075.
829. Pandurova, V., Nikolova, V. G., Stoichev, S. M., Arojo, M. I., Laskova, V. and Konstantinova, P. P., "Thin resistors of film NiCr obtained by the sublimation method", Elektro Prom.-st. & Priborostr., 12 (4), 134-6 (1977), Bulgarian, EEA81-14316.
830. Patz, U., "Sputtering technique for thin-film resistors and capacitors in microelectronics", Finommech-Mikrotech., 15 (11), 321-6 (1976), Hungarian, EEA80-35905.
831. Pedder, D. J., "Techniques for the examination of thick film resistor microstructures by TEM", Electrocompon. Sci. & Technol., 4 (2), 85-8 (1977), EEA81-5620.
832. Pike, G. E., "Conductivity of thick film (cermet) resistors as a function of metallic particle volume fraction", AIP Conf. Proc., no. 40, 366-71 (1978), EEA81-33021.

833. Pike, G. E. and Seager, C. H., "Electrical conduction mechanisms in thick film resistors", Report SAND-76-5329, Sandia Labs., Albuquerque, N. Mex., 1976, 13 pp.

834. Pike, G. E. and Seager, C. H., "Electrical properties and conduction mechanisms of Ru-based thick-film (cermet) resistors", J. Appl. Phys., $\underline{48}$ (12), 5152-69 (1977), EEA81-23632.

835. Polaschegg, H. D., Heinz, B. and Sommerkamp, P., "ESCA measurements on Ni-Cr resistor layers", Finommech. & Miktrotech., $\underline{15}$ (8), 236-9 (1976), Hungarian, EEA80-3296.

836. Polvorinos, J. J., "Resistor manufacturing technologies", Mundo Electron., no. 59, 45-52 (1977), Spanish, EEA80-16972.

837. Prinsen, P., "Thin resistive thick-film layers based on precious metal resinate/glass systems", Electrocompon. Sci. & Technol., $\underline{5}$ (1), 41-3 (1978), EEA81-37911.

838. Prudenziatai, M., "On the temperature coefficient of resistivity in thick-film resistors and the percolation model", Alta Freq., $\underline{46}$ (6), 287-93 (1977), EEA80-40990.

839. Prudenziatai, M. and Cattaneo, A., "Thermoelectric power in thick film resistors", Electrocompon. Sci. & Technol., $\underline{3}$ (3), 181-3 (1976), EEA80-17040.

840. Rhee, J. G. and Chen, T. M., "Contact noise in thick film resistors", Solid State Technol., $\underline{21}$ (9), 59-62 (1978), EEA82-643.

841. Ringo, J. A., Stevens, E. H. and Gilbert, D. A., "On the interpretation of noise in thick-film resistors", IEEE Trans. Parts, Hybrids & Packag., PHP-$\underline{12}$ (4), 378-80 (1976), EEA80-13285.

842. Robertson, J., "Conduction processes in high value thick film resistors", Electrocompon. Sci. & Technol., $\underline{4}$ (2), 105-9 (1977), EEA81-5622.

843. Rysanek, V., "The prediction of resistors stability with the aid of measured third harmonics", Acta Polytech. III, no. 4, 75-80 (1973), Czech, EEA79-20514.

844. Sakurai, S. and Doken, M., "Tantalum family thin film resistor", Patent Japan 75-40600, Publ. December 1975, CA84-158800.

845. Savitsky, E. M., Tylkina, M. A., Zhdanova, L. L., Kondratov, N. M., Marin, K. G. and Rogova, I. V., "Thin-film resistors of rhenium and rhenium-based alloys", J. Electron. Mater., $\underline{5}$ (4), 427-33 (1976), EEA80-17037.

846. Schaffer, C. A. and Sergent, J. E., "The effect of particle-size distribution on the electrical properties of $RuO_2$ thick-film resistors", Proceedings of the 1977 International Microelectronics Symposium, p. 60-4.

847. Schmitt, A., "Structuring and production of resistance networks in thin film technique", Elektron. J., $\underline{13}$ (1), 24-6 (1978), German, EEA81-42149.

848. Schuffny, R., "A new effective algorithm for the determination of AC and DC characteristics of spatially distributed resistive films", Nachrichtentech. Elektron., $\underline{27}$ (4), 155-7 (1977), German, EEA80-28554.

849. Seager, C. H. and Pike, G. E., "Electrical properties of DuPont birox and cermalloy thick film resistors - III. Electric field effects", Report SAND-76-0536, Sandia Labs., Albuquerque, N. Mex., 1976, 135 pp.

850. Sementsov, V. I. and Prozorovskiy, V. E., "Calculation of the parasitic parameters of film resistors in multilayer integrated structures", Telecommun. & Radio Eng. Pt. 2, 3 (8), 101-6 (1975), EEA79-37657.

851. Sergent, J. E. and Dryden, W. G., "Design method for thick film resistors", Solid State Technol., 20 (10), 63-7 (1977), EEA81-14330.

852. Sergent, J. E. and Koszykowski, A. W., "Designing thick film resistors", Electron. Prod., 6 (9), 48, 50, 53-4, 56, 59-60, 62, 64, 66 (1977), EEA81-19159.

853. Shah, J. S. and Berrin, L., "Mechanism and control of post-trim drift of laser-trimmed thick-film resistors", IEEE Trans. Components, Hybrids and Manuf. Technol., CHMT-1 (2), 130-6 (1978), EEA81-33025.

854. Shirakawa, T., Yoshida, S. and Murasumi, T., "Development of thin film circuits without a resistance trimming process", Fujitsu Sci. & Tech. J., 14 (1), 103-18 (1978), EEA81-33017.

855. Singh, A., "Techniques of adjusting thin- and thick-film resistors in hybrid microelectronic circuits", Microelectron. & Reliab., 15 (2), 123-9 (1976), EEA79-33157.

856. Singh, A., "Design and photoetch fabrication process for thin film resistor for hybrid ICs", J. Sci. & Ind. Res., 36 (1), 15-18 (1977), EEA81-10038.

857. Smyth, R. T. and Anderson, J. C., "Production of resistors by arc plasma spraying", Electrocompon. Sci. & Technol., 2 (2), 135-45 (1975), EEA79-4734.

858. Spiller, D. O., "The development of precision thin film resistors for submerged repeater applications", Electrocompon. Sci. & Technol., 5 (1), 9-14 (1978), EEA81-37903.

859. Starr, C. D. and Graule, R. S., "Precision secondary resistors", J. Test. & Eval., 5 (3), 210-16 (1977), EEA80-28553.

860. Stecher, G., "A new concept in calculation of thick film resistors", Electrocompon. Sci. & Technol., 4 (3-4), 163-9 (1977), EEA81-19157.

861. Stein, S. J., Huang, C. and Gelb, A. S., "A high voltage, high performance thick film resistor system", Electrocompon. Sci. & Technol., 4 (2), 95-104 (1977), EEA81-5621.

862. Stevens, E. H., Gilbert, D. A. and Ringo, J. A., "High-voltage damage and low-frequency noise in thick-film resistors", IEEE Trans. Parts, Hybrids & Packag., PHP-12 (4), 351-6 (1976), EEA80-13284.

863. Styblinski, M. A., "Some results of analytical study of resistor correlations in ICs based on the model of factor analysis", Proceedings of the 1978 IEEE International Symposium on Circuits and Systems, p. 205-6, EEA81-42445.

864. Sugita, E., Yasuda, T. and Matsumoto, T., "A solid-state vari-
     able resistor using a junction FET", IEEE Trans. Parts,
     Hybrids & Packag., PHP-12 (13), 260-4 (1976), EEA80-421.
865. Sutcliffe, H. and Ulgen, Y., "Spectra of AC-induced noise in
     resistors", Electron. Lett., 13 (14), 397-9 (1977), EEA80-35772.
866. Swart, P. L. and Van Wyck, J. D., "Resistor loops in hybrid
     circuits", Microelectronics, 7 (3), 53-5 (1976), EEA79-33159.
867. Sztaba, O. and Lusniak-Wojcicka, D., "Glass films as a pro-
     tection of thick film resistors", Pr. Przem. Inst. Elektron.,
     16 (1), 47-54 (1975), Polish, EEA81-14329.
868. Takano, T., "Radiation damage in components", Trans. Inst.
     Electron. & Commun. Eng. Jpn. Sect. E., E60 (5), 260-1 (1977),
     EEA81-5541.
869. Tetzner, K., "It won't work without thick film resistors",
     Funkschau, 50 (5), 179-80 (1978), German, EEA81-33024.
870. Thorbjornsen, A. R., "The statistical simulation of integrated
     resistors", Proceedings of the 1976 IEEE International Sym-
     posium on Circuits and Systems, p. 236-4, EEA80-3302.
871. Thorbjornsen, A. R., Dvorack, M. A. and Riad, A., "Correlation
     between thick-film resistance values", IEEE Trans. Parts,
     Hybrids & Packag., PHP-13 (2), 138-43 (1977), EEA80-32243.
872. Thorbjornsen, A. R. and Lee, S. I., "Estimation of correlation
     coefficients between film resistors", Proceedings of the 1978
     IEEE International Symposium on Circutis and Systems, p. 200-4,
     EEA81-42156.
873. Trick, T. N. and Chien, R. T., "Note on single fault detection
     in positive resistor circuits", IEEE Trans. Circuits Syst.,
     CAS-25 (1), 46-8 (1978).
874. Turczak, J. and Wtorkiewicz, A., "Thick film resistors screen
     printed on organic substrates with pastes baked at tempera-
     tures about 200°C", Pr. Przem. Inst. Elektron., 16 (1), 43-6
     (1975), Polish, EEA81-14328.
875. Ueno, F., Inoue, T. and Shirai, Y., "A floating optically
     controllable negative resistance element, its analysis and
     application", Trans. Inst. Electron. & Commun. Eng. Jpn.
     Sect. E, E60 (11), 693 (1977), EEA81-14174.
876. Ulbrich, W., "Resistor geometry comparison with respect to
     current noise and trim sensitivity", Electrocompon. Sci. &
     Technol., 4 (2), 63-8 (1977), EEA81-5613.
877. Vandamme, L. K. J., "Criteria of low-noise thick-film resis-
     tors", Electrocompon. Sci. & Technol., 4 (3-4), 171-7 (1977),
     EEA81-14324.
878. Van Hise, J. A., "Resistor on multilayer ceramic (MLC) en-
     hancement", IBM Tech. Disclosure Bull., 20 (11), 5170-1 (1978).
879. Von Stein, C., "Overglazing of resistors contemporary?",
     International Conference on Manufacturing and Packaging
     Techniques for Hybrid Circuits, 1976, p. 79-86, EEA79-46398.
880. Wahlers, R. L. and Merz, K. M., "High range low cost resistor
     system", Proceedings of the 28th Electronic Components Con-
     ference, 1978, p. 72-9, EEA81-37917.

881. Watanabe, K., Kawamata, S. and Kozaki, R., "High accuracy hybrid type 12-bit D/A converter made of TaAl-N thin film resistors", Proceedings of the 28th Electronic Components Conference, 1978, p. 253-9, EEA81-37927.
882. Weissman, M. B., "Reconciliation between thermodynamics and noise measurements on metal film resistors", Phys. Rev. Lett., 41 (1), 1-3 (1978), EEA81-37817.
883. Welsh, F. M., "Characteristics of low ohmic, high-power, metallized thick-film resistors", Proceedings of the 28th Electronic Components Conference, 1978, p. 58-71, EEA81-37916.
884. Wertz, N., "Conductive plastic thick film resistors", NTG-Fachber., no. 60, 40-4 (1977), German, EEA81-23637.
885. Wheatley, D. E., "Designing with precision resistors", New Electron., 11 (5), 114, 116 (1978), EEA81-28063.
886. Wheatley, D. and Bean, J., "The impact of bulk alloys on resistor technology", Electron, no. 114, 59, 66 (1977), EEA80-35777.
887. Willig, W. R., "Properties, processing and application of Ni, Cr and Ni-Cr for vacuum thin film technology", Finommech. & Mikrotech., 15 (8), 225-31 (1976), Hungarian, EEA80-3294.
888. Wilson, J. P., "What's happening to resistors?", New Electron., 11 (5), 110, 112 (1978), EEA81-28062.
889. Wood, J. H., "One percent thick-film resistors are made, not born", Proceedings of the 1977 International Microelectronics Symposium, p. 81-7.
890. Wood, J. H. "Control of pre-trim variables in thick film resistor processing", Solid State Technol., 21 (9), 67-72 (1978), EEA82-645.
891. Wu, T. S., Shiramatsu, T., Fu, S. L. and Liang, M. S., "Electrical properties of carbon black/resin thick film resistors", NTG-Fachber., no. 60, 34-9 (1977), German, EEA81-23636.
892. Yamazaki, J., Kamo, T. and Saito, Y., "Thin film resistor and a method of producing the same", Patent USA 4042479, Publ. August 1977
893. Yoshida, H., "Current-crowding effect on current noise of planar resistors", J. Appl. Phys., 49 (3), 1159-61 (1978), EEA81-32934.

X.   CAPACITORS

894. Adelson, L., "Production engineering measure for improved reliability of subminiature metallized polycarbonate capacitors", Report AD-A015529/1SL, Sprague Electric Co., North Adams, Mass., 1975, 249 pp.
895. Aglietti, G. and Rinaldi, M., "Criteria for design and performance control of a new series of dumping and commutation mixed dielectric capacitors", 2nd International Conference on Power Electronics-Power Semiconductors and their Applications, 1977, p. 33-6, EEA80-41008.

896. Ajit, C. N. and Jawalekar, S. R., "Thin film Al$_2$O$_3$ capacitors",
     Thin Solid Films, $\underline{37}$ (1), 85-9 (1976), EEA79-41584.
897. Albella, J. M. and Martinez-Duart, J. M., "Tantalum solid
     electrolytic capacitors. I. Anode characteristics", Electron.
     & Fis. Apl., $\underline{19}$ (2), 83-90 (1976), Spanish, EEA80-16977.
898. Albella, J. M. and Martinez-Duart, J. M., "Tantalum solid
     electrolytic capacitors. II. Properties of the dielectric
     (Ta$_2$O$_3$)", Electron. & Fis. Apl., $\underline{19}$ (3), 134-42 (1976),
     Spanish, EEA81-9958.
899. Albella, J. M. and Martinez-Duart, J. M., "Tantalum solid
     electrolytic capacitors. III. The cathode", Electron. &
     Fis. Apl., $\underline{19}$ (4), 185-9 (1976), Spanish, EEA81-32937.
900. Alexander, J. H., "Monolithic ceramic capacitors", Patent USA
     4089813, Publ. May 1978.
901. Amin, R. B., Anderson, H. U. and Hodgkins, C. E., "Low tempera-
     ture fired ceramic capacitors", Patent USA 4082906, Publ.
     April 1978.
902. Anon., "Screen-printable capacitor dielectrics", Circuits
     Manuf., $\underline{16}$ (9), 28, 30 (1976), EEA80-9196.
903. Anon., "Miniature metallised polyester film type capacitors",
     Electron. & Microelectron. Ind., no. 268, 46-7 (1976),
     EEA80-20542.
904. Anon., "Smaller capacitors through thinner foils", Elektronik,
     $\underline{26}$ (1), 58 (1977), German, EEA80-16978.
905. Anon., "Secure capacitor leads with tape during dipping and
     oven processing to maintain critical spacing", Insul./Circuits,
     $\underline{23}$ (1), 38-9 (1977), EEA80-9132.
906. Anon., "Take a close look at the '77 caps", Eval. Eng., $\underline{16}$
     (3), 36-7, 39, 42, 44 (1977), EEA80-35785.
907. Anon., "The tantalum electrolytic capacitor", Funkschau, $\underline{49}$
     (18), 841-2 (1977), German, EEA81-19089.
908. Anon., "Capacitors", Can. Electron. Eng., $\underline{21}$ (10), 34 (1977),
     EEA81-19082.
909. Anon., "Beam tape aids capacitor fabrication", Electron.
     Packag. & Prod., $\underline{18}$ (10), 117-18, 120 (1978).
910. Ansell, J. L., Brusius, P. G. and Baker, R. J., "Method for
     making thick film capacitors", Patent USA 4007296, Publ.
     February 1977.
911. Arnold, A. J., Hardin, P. W. and Rippen, R. A., "Removable
     decoupling capacitor", IBM Tech. Disclosure Bull., $\underline{21}$ (5),
     1856 (1978).
912. Arnold, A. J., Rippens, R. A. and Woodside D. H., "Incorpora-
     tion of decoupling capacitors into connector", IBM Tech. Dis-
     closure Bull., $\underline{20}$ (11), 4338 (1978).
913. Artbauer, O., "Stable low value capacitors", European Confer-
     ence on Precise Electrical Measurement, 1977, p. 56-7,
     EEA80-41006.
914. Arvanitakis, N. C., Jones, N. G., Stevens, K. T. and Winkler,
     P. E., "Symmetrical design for polarized capacitor", IBM Tech.
     Disclosure Bull., $\underline{21}$ (4), 1407 (1978).

915. Atkinson, A., "Capacitor selection by circuit function", Electron. Ind., 3 (5), 12-13 (1977), EEA80-24541.
916. Azzis, D., "Composite integrated capacitor", IBM Tech. Disclosure Bull., 20 (7), 2714-15 (1977), EEA81-46053.
917. Bailey, R. A. and Nevin, J. H., "Thin-film multilayer capacitors using pyrolytically deposited silicon dioxide", IEEE Trans. Parts, Hybrids & Packag., PHP-12 (4), 361-4 (1976), EEA80-13282.
918. Balakleev, V. N. and Zolotarev, N. A., "Analysis of edge zone partial capacities of a plane capacitor with a protective ring", Izv. VUZ Elektromekh., no. 6, 628-36 (1977), Russian, EEA80-35790.
919. Baumann, M., "The self-sealing effect in metallised moulded condensers", Bauelem. Elektrotech., 12 (3), 42-50 (1977), German, EEA80-41000.
920. Bender, L., "Non PCB electrolytics [polychlorinated biphenyl]", Can. Electron. Eng., 21 (10), 35-6 (1977), EEA81-14177.
921. Bentley, P., "Choosing monolithic multi-layer ceramics [for capacitors]", Electron, no. 121, 55, 57 (1977), EEA81-19086.
922. Bernard, W. J., "Developments in electrolytic capacitors", J. Electrochem. Soc., 124 (12), 403C-9C (1977), EEA81-14185.
923. Bieger, F., "MKV power capacitors", Siemens Rev., 44 (1), 44-6 (1977), EEA80-32184.
924. Binner, H., "Multilayer circuits with thin film capacitors", Elektron. Anz., 8 (12), 297-300 (1976), German, EEA80-9193.
925. Borough, J. W., Burnham, J., Simmons, W. J. and Webster, S. L., "Spurious signal generation in plastic film capacitors", IEEE Trans. Parts, Hybrids & Packag., PHP-13 (4), 402-6 (1977), EEA81-14184.
926. Bowling, E. L. and Papadopoulos, G. S., "A choice of electrolytic capacitors: aluminium or tantalum?", Bauelem. Elektrotech., 11 (11), 36, 38, 43-4, 46 (1976), German, EEA80-16976.
927. Bratschun, W. R., "Glass-passivated thick-film capacitors for RC circuits", IEEE Trans. Parts, Hybrids & Packag., PHP-12 (13), 194-201 (1976), EEA80-500.
928. Brennan, T. F., "Ceramic capacitor insulation resistance failures accelerated by low voltage", IEEE Trans. Electron. Devices, ED-26 (1), 102-8 (1979).
929. Brettle, J. and Jackson, N. F., "Failure mechanisms in solid electrolytic capacitors", Electrocompon. Sci. & Technol., 3 (4), 233-46 (1977), EEA80-35787.
930. Bright, J., "Examine axial ceramic capacitors with auto insertion in mind", EDN, 22 (20), 93 (1977), EEA81-19088.
931. Brown, W. D. and Grannemann, W. W., "Metal-titanium dioxide-silicon capacitors", 1974 International Electronics Devices Meeting, Technical Digest, p. 566-8.
932. Buchanan, J. E., "Dielectric absorption - it can be a real problem in timing circuits", EDN, 22 (2), 83-6 (1977), EEA80-24542.

933. Buchanan, J. E., "Recovery voltage measurements with emphasis below one second for Class I NPO ceramic capacitors", IEEE Trans. Components, Hybrids & Manuf. Technol., CHMT-$\underline{1}$ (1), 112-14 (1978), EEA81-32940.

934. Burn, I., "Ceramic disk capacitors with base-metal electrodes", Am. Ceram. Soc. Bull., $\underline{57}$ (6), 600-1, 604 (1978).

935. Burnham, J., Webster, S. L., Simmons, W. J. and Borough, J. W., "A study of dielectric absorption in capacitors by thermally stimulated discharge (TSD) tests", 14th Annual Proceedings Reliability Physics Symposium, 1976, p. 147-56, EEA80-16981.

936. Burough, J. W., Brammer, W. G. and Burnham, J., "Degradation of PVF$_2$ capacitors during accelerated tests", 16th Annual Proceedings Reliability Physics Symposium, 1978, p. 219-23.

937. Burrow, F. H., "Trimmer capacitors: a guide to their selection", Electron, no. 103, 66, 68 (1976), EEA80-425.

938. Carlos, M. F. and Jon, M. C., "Detection of cracking during rotational soldering of a high reliability and voltage ceramic capacitor", Proceedings of the 28th Electronic Components Conference, 1978, p. 336-9, EEA81-42089.

939. Carpenter, A., "Factors which influence capacitor choice", New Electron., $\underline{11}$ (5), 139, 141 (1978), EEA81-28077.

940. Carr, L. A., Jr. and Janton, W. M., "Stress relief of metal dielectric metal capacitors", RCA Tech. Not., no. TN 1164, 1-2 (1976), EEA80-9134.

941. Caves, J. T., Copeland, M. A., Rahim, C. F. and Rosenbaum, S. D., "Sampled analog filtering using switched capacitors as resistor equivalents", IEEE J. Solid State Circuits, SC-$\underline{12}$ (6), 592-99 (1977).

942. Cernetic, J., "Liquid electrolyte for the use in aluminium-foil electrolytic capacitors for the temperature range of -55°C to +125°C", Elektroteh. Vestn., $\underline{44}$ (3), 181-6 (1977), Slovene, EEA81-19085.

943. Chatinyan, Y. S., "Determining overheating in power capacitors under non-steady-state operating conditions", Izv. VUZ Energ., no. 9, 37-41 (1976), Russian, EEA80-426.

944. Chernyaev, P. A., "Investigating absorption properties of fluoroplastic film capacitors", Meas. Tech., $\underline{9}$ (19), 1322-3 (1976), EEA80-41005.

945. Chlopek, J. and Przybysz, C., "The parameters of small capacity monolithic capacitors in UHF band to 1 GHz", Elektronika, $\underline{18}$ (7-8), 276-8 (1977), Polish, EEA81-5543.

946. Choudhury, T. K. D. and Bhattacharjee, P. K., "Electrolytic capacitor", Electr. India, $\underline{18}$ (3), 25-8 (1978), EEA81-42082.

947. Closa, C., "AF and power ceramic capacitors", Metal. & Electr., $\underline{41}$ (472), 120-1, 123 (1977), Spanish, EEA80-16980.

948. Coda, N. and Selvaggi, J. A., "Design considerations for high-frequency ceramic chip capacitors (for hybrid ICs)", IEEE Trans. Parts, Hybrids & Packag., PHP-$\underline{12}$ (13), 206-12 (1976), EEA80-427.

949. Cowdell, R. B., "DC to daylight filters", International Symposium on Electromagnetic Compatibility, 1976, p. 136-41, EEA80-3254.

950. Cozzolino, M. J., Galvagni, J. L. and Ewell, G. J., "Nondestructive examination of multilayer capacitors by neutron radiography", IEEE Trans. Components, Hybrids & Manuf. Technol., CHMT-1 (3), 265-73 (1978).

951. Crownover, J. W., "A low firing K3000 (X7R) dielectric", Proceedings of the 28th Electronic Components Conference, 1978, p. 204-6, EEA81-37830.

952. Dale-Lace, J. D., "High reliability electrolytic capacitors", New Electron., 10 (10), 128, 131 (1977), EEA81-5547.

953. Dale-Lace, J., "The capacitor to suit the job", New Electron., 11 (5), 131, 133 (1978), EEA81-28074.

954. Dalton, R. P., "Usage constraints for tantalum foil capacitors", Eval. Eng., 16 (2), 28-9 (1977), EEA80-41002.

955. Dance, M., "Capacitor developments", Electron. Ind., 4 (6), 37-41, 43 (1978), EEA81-37822.

956. Davidson, E. E., Katopis, G. A., Parisi, J. A. and Rubin, B. J., "Capacitor for multichip modules", IBM Tech. Disclosure Bull., 20 (8), 3117-8 (1978).

957. De LaMoneda, F. H., "Charge pumping device with integrated regulating capacitor and method for making same", Patent USA 4115794, Publ. September 1978.

958. De Mey, G. and De Wilde, W., "The influence of the electronic current on the measurement of ion polarisation in evaporated silicon oxide", Thin Solid Films, 42 (1), 81-9 (1977), EEA80-28558.

959. DeMey, G. and Pauwels, H. J., "Time constant of the ion current in thin film capacitors", Phys. Status Solidi A, 44 (2), K195-9 (1977), EEA81-14186.

960. De Wilde, W. and De Mey, G., "Study of low-frequency relaxation phenomena in thin-film capacitors", Appl. Sci. Res., 31 (6), 401-30 (1976), EEA79-24897.

961. Doken, M., Ohwada, K., Okamoto, S. and Kamei, T., "Thin-film capacitors made from TaN films", IEEE Trans. Components, Hybrids & Manuf. Technol., CHMT-1 (2), 187-91 (1978), EEA81-33020.

962. Domingos, H., Quattro, D. P. and Scaturro, J., "Breakdown in ceramic capacitors under pulsed high-voltage stress", IEEE Trans. Components, Hybrids & Manuf. Technol., CHMT-1 (4), 423-8 (1978).

963. Driver, P. D., "The development of the capacitor industry and its position in world markets", Electrocompon. Sci. & Technol., 5 (2), 119-26 (1978), EEA81-42085.

964. Edge, J., "New applications for metallized film capacitors", SME Tech. Pap. Ser. EE, Paper EE76-568, 1976, 9 pp.

965. Ehrich, M., "A contribution to the calculation of cylindrical capacitors", Arch. Elektrotech., 58 (3), 129-33 (1976), German, EEA80-424.

966. England, W. F., "Tantalum-cased wet-slug tantalum capacitors",
     27th Electronic Components Conference, 1977, p. 379-86,
     EEA80-35797.
967. Epand, D. and Liddane, K., "Selecting capacitors properly",
     Electron. Des., 25 (13), 66-71 (1977), EEA81-9959.
968. Ermuratskii, V. V. and Ermuratskii, P. V., "Calculating the
     heat production in capacitors on applying standard forms of
     voltage", Elektrichestvo, no. 11, 45-51 (1976), Russian,
     EEA80-41001.
969. Ewell, G. J. and Jones, W. K., "Encapsulation, sectioning,
     and examination of multilayer ceramic chip capacitors", 27th
     Electronic Components Conference, 1977, p. 446-51, EEA80-35800.
970. Finger, B., "Structure and advantages of MKH layer capacitors",
     Components Rep., 12 (2), 34-7 (1977), EEA81-19083.
971. Frederick Hanold, R. C., III, "Low temperature sintering ce-
     ramic capacitor and method of making", Patent USA 4081857,
     Publ. March 1978.
972. French, G. A., "Electrolytic capacitance meter", Radio &
     Electron. Constructor, 30, 411-14 (1977), EEA80-24545.
973. Fritzsche, H., "Capacitances of coplanar microstrip lines in
     integrated circuits", Siemens Forsch.- & Entwicklungsber., 5
     (2), 72-5 (1976), EEA79-24927.
974. Fujinoki, Y., "MP capacitors: used for special requirements",
     JEE, no. 133, 22-6 (1978), EEA81-37824.
975. Gaenge, F., "No breakdown even at overvoltages", Funk-Tech.,
     31 (17), 522-6 (1976), German, EEA80-9131.
976. Gaenge, F. and Gottlob, H., "Test method for assessing the
     safety in operation of self-healing (MK) capacitors", Energie,
     28 (12), 366-8 (1976), EEA80-9130.
977. Gagvani, P. H., "Use of polypropylene film in power capacitors",
     Electr. India, 18 (3), 5-8 (1978), EEA81-42078.
978. Garrett, K., "Tantalum capacitor technqiue", New Electron.,
     11 (5), 138-9 (1978), EEA81-28076.
979. Geresi, K., "Present state and trends in the development of
     electrolytes", Hiradastechnika, 28 (8), 233-6 (1977),
     Hungarian, EEA81-19090.
980. Girard, R. T. and Rice, G. A., "Thick film capacitor", Patent
     USA 3968412, Publ. July 1976.
981. Girard, R. T. and Rice, G. A., "High dielectric constant ink for
     thick film capacitors", Patent USA 4061584, Publ. December 1977.
982. Gmitrzak, A. and Piotrowski, J., "Some aspects of high voltage
     power capacitors testing", Pr. Inst. Elektrotech., 24 (95),
     152-6 (1976), Polish, EEA80-6388.
983. Goffaux, R., "The parameters responsible for the demetallisa-
     tion of self-healing low-voltage condensers", Bull. Soc. R.
     Belge Electr., 92 (4), 221 (1976), French, EEA80-16975.
984. Goldman, E. I., Gulyaev, I. B., Zhdan, A. G. and Messerer, M.
     A., "Thermally stimulated and isothermal discharges of a capac-
     itor across a semiconductor with grain-boundary barriers",
     Sov. Phys., Semicond., 11 (3), 300-3 (1977).

985. Goudswaard, B. and Driesens, F. J. J., "Failure mechanism of
     solid tantalum capacitors", Electrocompon. Sci. & Technol., 3
     (3), 171-9 (1976), EEA80-16979.
986. Harrop, P. J., "Recent advances in capacitors", Electron. Ind.,
     3 (4), 15-17, 19 (1977), EEA80-24540.
987. Harrop, P. J., "Capacitor new manufacturing technologies",
     Mundo Electron., no. 65, 57-61 (1977), Spanish, EEA81-9964.
988. Hart, M. L. and Minetti, R. H., "Method of grading mica for
     use in high-reliability capacitors", IEEE Trans. Parts, Hybrids
     & Packag., PHP-12 (2), 105-10 (1976).
989. Hayashi, T. and Onuki, M., "Thin-film capacitor and method for
     the fabrication thereof", Patent Canada 982666, Publ. January
     1976.
990. Helwig, G., "Aluminum electrolytic condenser to DIN 41259 in
     moulded cases", Bauelem. Elektrotech., 12 (3), 32, 34, 36, 38
     (1977), German, EEA80-40999.
991. Hirata, K. and Yamasaki, T., "Miniaturized aluminum solid
     electrolytic capacitors using a highly effective surface en-
     largement technique", IEEE Trans. Parts, Hybrids & Packag.,
     PHP-12 (13), 271-23 (1976), EEA80-428.
992. Holladay, A. M., "All-tantalum wet-slug capacitor overcomes
     catastrophic failure", Electronics, 51 (4), 105-8 (1978),
     EEA81-19084.
993. Holladay, A. M., "Guidelines for the selection and application
     of tantalum electrolytic capacitors in highly reliable equip-
     ment", Report N78-18311/8SL, NASA, Marshall Space Flight
     Center, Huntsville, Ala., 1978, 68 pp.
994. Holland, H. W., "Inspection of Class 2 ceramic capacitors",
     Eval. Eng., 17 (1), 26-8 (1978), EEA81-37821.
995. Hrovat, M., Stadler, Z. and Kolar, D., "Dielectric materials
     for thick film capacitors", Elektroteh. Vestn., 44 (3),
     167-71 (1977), Slovene, EEA81-19155.
996. Hudoklin, A. and Kavcic, P., "Estimation of reliability of
     capacitors using an automatic system", Elektroteh. Vestn.,
     44 (4), 251-5 (1977), Slovene, EEA81-28067.
997. Inoue, J., "Ceramic capacitors must meet severe standards",
     JEE, no. 133, 27-30 (1978), EEA81-37825.
998. Irlin, A. V., Vlasova, V. V. and Fabrikant, B. A., "Character-
     istics and design considerations of film capacitors with di-
     electrics using oxides of rare earth metals", Poluprovodn.
     Tekh. & Mikroelektron., no. 24, 80-2 (1976), Russian, EEA80-492.
999. Itoh, Y. and Yoshimura, S., "Alternating-current properties
     of organic semiconductor solid capacitors", J. Electrochem.
     Soc., 124 (7), 1128-33 (1977), EEA80-41004.
1000. Jostel, J., "Practical results using topographic methods to
      determine reliability of ceramic capacitors", Radio Fern-
      sehen Elektron., 25 (23), 749-52 (1976), German, EEA80-24544.
1001. Juergens, W., "Voltage drop and charge consumption of anodized
      tantalum films", NTG-Fachber., no. 60, 161-6 (1977), German,
      EEA81-23627.

1002. Kagawa, C., Miyamoto, A. and Miyajima, K., "Aluminum foils for electrolytic capacitors", Patent Japan 78-19151, Publ. February 1978, CA89-119712.

1003. Kakkar, S. S., "Field failures in capacitors", Electr. India, 18 (3), 13-15 (1978), EEA81-42080.

1004. Kashiwazaki, T., Asakawa, J. and Akiyama, K., "Toshiba power capacitor impregnated with non-PCB oil (non-polychlorinated diphenyl oil)", Toshiba Rev. (Int. Ed.), no. 108, 9-12 (1977), EEA80-35795.

1005. Kilgenstein, O., "Components - fundamentals for practical people. V (capacitors)", Funk-Tech., 32 (8), 87-93 (1977), German, EEA80-35789.

1006. Kinugawa, K., "Recent capacitor techniques", JEE, no. 125, 56-9 (1977), EEA81-23527.

1007. Kinukawa, K., "Capacitor technology in Japan", JEE, no. 133, 18-21 (1978), EEA81-37823.

1008. Klonz, M. and Waldele, F., "Two-plate capacitor with aerostatic electrode regulation for the precise determination of capacitor loss factor", Elektrotech. Z. ETZ B, 30 (5), 171-2 (1978), German, EEA81-42077.

1009. Kobayashi, T., Ariyoshi, H. and Masuda, A., "Reliability evaluation and failure analysis for multilayer ceramic chip capacitors", IEEE Trans. Components, Hybrids & Manuf. Technol., CHMT-1 (3), 316-24 (1978).

1010. Konotop, V. V. and Rydakov, V. V., "On calculating the electric field at the edge of a plane capacitor having a combined dielectric", Izv. VUZ Energ., no. 5, 132-7 (1977), Russian, EEA80-32181.

1011. Kurumizawa, H., "Aluminum electrolytic capacitors for consumer electronics", JEI-Jpn. Electron. Ind., 24 (5), 33-5 (1977), EEA81-9962.

1012. Leigh, W. C., "An incrementally adjustable fixed ceramic chip capacitor", Proceedings of the 1977 International Microelectronics Symposium, p. 253-7.

1013. Lerner, M. M., "Calculating the losses in paper-dielectric capacitors under non-sinusoidal low-frequency voltage", Sov. Electr. Eng., 47 (2), 90-4 (1976), EEA80-35794.

1014. Lerner, M., "Which non-sinusoidal voltage may shorten the life of a capacitor?", 27th Electronic Components Conference, 1977, p. 468-74, EEA80-35803.

1015. Lerner, M., "When heat losses in a better (low dissipation factor) capacitor may be more than in a poorer one (non-sinusoidal voltage waveforms)", IEEE Trans. Parts, Hybrids & Packag., PHP-13 (2), 166-73 (1977), EEA80-32182.

1016. Lerner, M., Place, L. and Hutzler, J. R., "Comparative design of non-PCB capacitors", 27th Electronic Components Conference, 1977, p. 475-80, EEA80-35804.

1017. Leroy, Y., "Descamps, M. and Vernet, M., "Characterization of thick film capacitors", Electrocompon. Sci. & Technol., 4 (3-4), 125-31 (1977), EEA81-14323.

1018. Licznerski, B., Nitsch, K. and Rzasa, B., "Low frequency dis-
      persion in thick-film capacitors with rutile-glass insula-
      tion", Arch. Elektrotech., 26 (4), 743-68 (1977), Polish,
      EEA81-37908.
1019. Lindquist, C. S., "Capacitor DF and Q", IEEE Trans. Compo-
      nents, Hybrids & Manuf. Technol., CHMT-1 (1), 115-17 (1978),
      EEA81-32941.
1020. Lipchinskii, A. G., "Line capacitance of a ferroelectric film
      planar capacitor", Izv. VUZ Radioelektron., 20 (7), 124-7
      (1977), Russian, EEA80-41003.
1021. Loesch, M. D., "Capacitor packaging technique", IBM Tech. Dis-
      closure Bull., 21 (3), 967 (1978).
1022. Loescher, D. H. and Gumley, C. E., "Surface arcover in DC
      capacitors", IEEE Trans. Parts, Hybrids & Packag., PHP-12
      (2), 97-99 (1976).
1023. Loescher, D. H. and Gumley, C. E., "How heat treatment affects
      film foil capacitors", IEEE Trans. Parts, Hybrids & Packag.,
      PHP-12 (13), 251-4 (1976), EEA80-429.
1024. Loescher, D. H. and Sidnell, N. A., "Film capacitors with
      low internal inductance", IEEE Trans. Parts, Hybrids &
      Packag., PHP-13 (4), 399-402 (1977), EEA81-9961.
1025. Longland, T., "Power capacitor dielectrics return to use of
      oil impregnants", Electr. Times, no. 4406, 5-6 (1976),
      EEA80-6385.
1026. Love, G. R. and Ewell, G. J., "Acoustic microscopy of ceramic
      capacitors", IEEE Trans. Components, Hybrids & Manuf. Technol.,
      CHMT-1 (3), 251-7 (1978).
1027. Love, G. R., McLaurin, E. D. and Hucks, W. E., "Tab lead
      capacitor", IEEE Trans. Parts, Hybrids & Packag., PHP-13 (3),
      279-82 (1977), EEA81-912.
1028. Love, W., III and Rosenberg, M., "Hermetic glass encapsulated
      capacitors", 27th Electronic Components Conference, 1977,
      p. 387-90, EEA80-35798.
1029. McClamrock, G. L., "Apparatus for electrical and heat stress-
      ing of radially-leaded capacitors", Tech. Dig., 48, 19-20
      (1977), EEA81-45676.
1030. Maher, G. H., "Improved dielectrics for multi-layer ceramic
      capacitors", 27th Electronic Components Conference, 1977,
      p. 391-9, EEA80-35799.
1031. Maher, G. H. and Burks, D. P., "High-voltage multi-layer cera-
      mic chip capacitors", Proceedings of the 1977 International
      Microelectronics Symposium, p. 258-61.
1032. Maher, J. P., Jacobsen, R. T. and Lafferty, R. E., "High-
      frequency measurement of Q-factors of ceramic chip capacitors",
      IEEE Trans. Components, Hybrids & Manuf. Technol., CHMT-1
      (3), 257-64 (1978).
1033. Mash, S., "Everything with chips (monolithic)", Electron.
      Equip. News, 26-7 (1976), EEA79-20527.

1034. Masumura, H., Fujiwara, S. and Tanaka, H., "Reduction-reoxidation type semiconductor ceramic capacitor", Patent USA 4073846, Publ. February 1978.

1035. Mayer, G. and Houska, K. H., "SiO capacitors for high frequency thin film circuits", NTZ Nachr. Z. NTZ Commun. J., 28 (6), 207-11 (1975).

1036. Miersch, E. F. and Spampinato, D. P., "Two device memory cell with single floating capacitor", Patent USA 4103342, Publ. July 1978.

1037. Millard, R. J. and Cheseldine, D. M., "A new tantalum chip capacitor", Proceedings of the 28th Electronic Components Conference, 1978, p. 422-6, EEA81-42091.

1038. Minagawa, T., "Plastic film capacitors keep up with new applications", JEE, no. 133, 31-5 (1978), EEA81-37826.

1039. Mishin, Y. S., "An algorithm for calculating the capacitance of topological structure elements in microcircuit films", Izv. VUZ Priborostr., 19 (11), 90-2 (1976), Russian, EEA80-9188.

1040. Mishin, Y. S., "Capacitance of film microcircuit components", Izv. VUZ Radioelektron., 20 (5), 126-8 (1977), Russian, EEA80-41107.

1041. Nidera, A., "The electrochemical processes in the manufacturing of electrolytic capacitors", Elektronika, 18 (1), 16-20, 29-30 (1977), Polish, EEA80-20541.

1042. Noorlander, W., "Some aspects of multilayer ceramic chip capacitors for hybrid circuits", Electrocompon. & Sci. & Technol., 5 (1), 33-40 (1978), EEA81-37919.

1043. Nordhage, F. and Backstrom, G., "Injection charging of polymeric capacitors", J. Electrostat., 2 (4), 317-26 (1977), EEA80-35792.

1044. Ohwada, K. and Sakuma, K., "Lowering the loss of the TMM capacitor", Rev. Electr. Commun. Lab., 25 (3-4), 217-23 (1977), EEA81-858.

1045. Ohwada, K. and Sakuma, K., "Characteristic analysis of TMM capacitor", Trans. Inst. Electron. & Commun. Eng. Jpn. Sect. E, E61 (3), 259-60 (1978), EEA81-42087.

1046. Ohwada, K., Sakuma, K., Kiuchi, K. and Tomimuro, H., "SiO$_2$ thin-film capacitor prepared by RF sputtering", Rev. Electr. Commun. Lab., 24 (1-2), 128-39 (1976), EEA79-24898.

1047. Paasikallio, S., "Developments in capacitor dielectrics", Electr. India, 18 (3), 9-12 (1978), EEA81-42079.

1048. Parker, R. D., "Effect of foil edge modifications and configurational changes on energy storage capacitor weight", IEEE Trans. Parts, Hybrids & Packag., PHP-13 (3), 314-17 (1977).

1049. Pedersen, O., "Ceramic capacitors", Elektronik, no. 8, 15-18 (1977), Danish, EEA81-14179.

1050. Pennebaker, W. B., "RF sputtered strontium titanate films", IBM J. Res. & Develop., 13 (6), 686-95 (1969).

1051. Pennington, S., Dillinger, T. and Guckel, H., "Properties of electromechanical devices utilizing thin silicon diaphragms", Proceedings of the 28th Electronic Components Conference, 1978, p. 435-9, EEA81-42092.

1052. Perinati, A., "Tantalum thin film applications - a new approach for capacitors", Electrocompon. Sci. & Technol., 4 (2), 69-73 (1977), EEA81-5614.

1053. Peters, F. G. and Schwartz, N., "Reverse bias life test stability of tantalum-titanium anodic oxide thin film capacitors", J. Electrochem. Soc., 124 (6), 949-51 (1977), EEA80-32183.

1054. Poole, R. B., "An improved aluminum electrolytic capacitor for timing circuit applications", Proceedings of the 28th Electronic Components Conference, 1978, p. 415-21, EEA81-42090.

1055. Potzlberger, H. W., "Thin film integrated RC-networks with compensated temperature coefficients of R and C", Electrocompon. Sci. & Technol., 4 (3-4), 139-42 (1977), EEA81-14317.

1056. Rao, M. K. and Jawalekar, S. R., "Improved thin film $Al_2O_3$ capacitors", Thin Solid Films, 51 (2), 185-8 (1978), EEA81-42086.

1057. Rau, S., Manafu, M. and Avram, E., "The influence of the oxidation voltage on the electrical characteristics of the $Ta_2O_5$ dielectric from the Ta electrolytic capacitors with synthesized anode", Electroteh. Electron. & Autom. Autom. & Electron., 21 (3), 107-11 (1977), Rumanian, EEA81-14180.

1058. Riekeles, R., "N-doped Ta thin film capacitors with improved performance", NTG-Fachber., no. 60, 155-60 (1977), German, EEA81-23626.

1059. Robbins, W. L., Jones, W. K. and Ewell, G. J., "The piezoelectric behavior of multilayer monolithic ceramic capacitors", Proceedings of the 1977 International Microelectronics Symposium, p. 235-40.

1060. Rottersman, M. H., Bill, M. J. and Gerstenberg, D., "Tantalum film capacitors with improved AC properties", 27th Electronic Components Conference, 1977, p. 462-7, EEA80-35802.

1061. Rottersman, M. H., Bill, M. J. and Gerstenberg, D., "Tantalum film capacitors with improved AC properties", IEEE Trans. Components, Hybrids & Manuf. Technol., CHMT-1 (2), 137-42 (1978), EEA81-33018.

1062. Ruchi, A. F. and Brennan, P. A., "Capacitance models for integrated circuit metallization wires", IEEE J. Solid-State Circuits, SC-10 (6), 530-6 (1975), EEA79-4752.

1063. Rybas, K. P. and Telepaev, B. N., "Modulation mechanism for an electron beam in a cylindrical capacitor", Sov. Phys.-Tech. Phys., 21 (8), 970-2 (1976), EEA80-24546.

1064. Rzasa, B., "Ageing and regeneration of thick film capacitors", Elektronika, 17 (5), 175-9 (1976), Polish, EEA79-41578.

1065. Sathe, G. V., "Equipment for capacitor manufacturing", Electr. India, 18 (3), 43-6 (1978), EEA81-42084.

1066. Sato, S., Yanagisawa, K., Okamoto, S. and Sasaki, H., "Therm-
      ally stable tantalum oxide thin film capacitors", 27th Elec-
      tronic Components Conference, 1977, p. 374-8, EEA80-35796.
1067. Schreier, P. G., "Capacitors-broad applications dictate a
      wide variety of devices", EDN, 22 (17), 46-56 (1977),
      EEA81-856.
1068. Seeba, M. D. and Sears, R. A., "Ceramic-chip-capacitor attach-
      ment", IEEE Trans. Parts, Hybrids & Packag., PHP-13 (4),
      395-9 (1977), EEA81-14183.
1069. Semin, A. S., Minnigulov, I. A., Dyachkov, V. I. and Koro-
      beinikov, P. V., "A film trimmer capacitor for hybrid micro-
      circuits", Izv. VUZ Radioelektron., 21 (1), 105-8 (1978),
      Russian, EEA81-28182.
1070. Semin, A. S., Suslov, Y. M. and Kravchenko, G. A., "Study of
      the film capacitors in the frequency range 200-1200 MHz",
      Izv. VUZ Radioelektron., 18 (12), 100-3 (1975), Russian,
      EEA79-13005.
1071. Shiomi, H. and Harada, T., "On correlation coefficients of
      degradation parameters in the electrolytic capacitors", Bull.
      Electrotech. Lab., 41 (9), 713-28 (1977), Japanese, EEA81-14176.
1072. Shirn, G. A., "Heating in polyethylene terephthalate film AC
      capacitors", IEEE Trans. Parts, Hybrids & Packag., PHP-12
      (2), 100-5 (1976).
1073. Shirn, G. A., Rice, H. L. and Linzey, R., "Corona starting
      voltage of capacitors driven with pulsed waveforms", IEEE
      Trans. Consum. Electron., CE-23 (1), 120-7 (1977), EEA80-20543.
1074. Singh, A., "Integration of capacitors into thick film hybrid
      integrated circuits", Stud. J. Inst. Electron. & Telecommun.
      Eng., 17 (4), 199-202 (1976), EEA80-41112.
1075. Singh, A., "Capacitors of thin insulating films of photo-
      resist materials", Microelectron. & Reliab., 17 (4), 441-4
      (1978), EEA81-45675.
1076. Singh, B. R., "Application of metal-insulator-metal (MIM)
      capacitors as a DC block in microwave integrated circuits",
      Thin Solid Films, 42 (3), L5-8 (1977), EEA80-32242.
1077. Smith, L. T., Apodaca, L. DeMartino, V. R. and Trew, J. W.,
      "Charge loss and recovery characteristics of irradiated tan-
      talum capacitors", IEEE Trans. Nucl. Sci., NS-24 (6), 2230-5
      (1977), EEA81-19091.
1078. Smithson, R. V., "Checking electrolytic capacitors", Radio &
      Electron. Constructor, 31 (5), 286-7 (1978), EEA81-19093.
1079. Sparkes, R., "Chip capacitor marking", Electron. Equip., 15
      (1976), EEA79-20559.
1080. Sprague Electric Co., "Temperature stable monolithic ceramic
      capacitor", Patent UK 1502558, Publ. March 1978.
1081. Spriggs, R. S. and Cronshagen, A. H., "Nondestructive, X-ray
      inspection of ceramic-chip capacitors for delaminations",
      14th Annual Proceedings Reliability Physics Symposium, 1976,
      p. 157-63, EEA80-16982.

1082. Spriggs, R. S. and Cronshagen, A. H., "X-rays provide non-destructive means for detecting chip capacitor delaminations", Insul./Circuits, 24 (2), 59-62 (1978), EEA81-28071.

1083. Sproull, J. F., Bacher, R. J., Larry, J. R. and Cote, R. E., "A high performance thick film capacitor system", Proceedings of the 28th Electronic Components Conference, 1978, p. 38-46, EEA81-37915.

1084. Sproull, J. F., Bacher, R. J., Larry, J. R. and Cote, R. E., "High K thick-film capacitors", Circuits Manuf., 18 (8), 27-8, 30, 32, 34 (1978).

1085. Srinivasan, S., "Tantalum capacitors", Electr. India, 18 (3), 37-42 (1978), EEA81-42083.

1086. Standard Telephone & Cables, "Monolithic ceramic capacitors", Patent UK 1478809, Publ. July 1977.

1087. Sunda, J. A., "Improved [capacitor] designs and better productivity", Electron, no. 121, 58 (1977), EEA81-19087.

1088. Sunda, J., "Multi-layer ceramic capacitors", New Electron., 11 (5), 142, 146, 148 (1978), EEA81-28078.

1089. Sutton, R. T., "The temperature coefficient of circuits containing several capacitors", Electron. Ind., 3 (7-8), 16-17 (1977), EEA80-35788.

1090. Swan, M. J. and Rayner, G. H., "A picofarad capacitor formed in a fused silica cavity", European Conference on Precise Electrical Measurement, 1977, p. 58-60, EEA80-41007.

1091. Tagare, D. M., "Is testing capacitors as per IS-2834 completely satisfactory?", Electr. India, 18 (3), 17-20 (1978), EEA81-42081.

1092. Takashima, K., "Low-loss, light-weight capacitor uses metallized plastic films on both surfaces", JEE, no. 119, 40-2 (1976), EEA80-20545.

1093. Tanabe, T., "Thin plastic films cut down size of capacitors", JEE, no. 119, 36-9 (1976), EEA80-25044.

1094. Tarr, M., "The passive components of thick film circuits", New Electron., 10 (6), 90-2 (1972), EEA80-35916.

1095. Tashiro, S., "Solid tantalum capacitors meet extreme requirements", JEE, no. 133, 36-40 (1978), EEA81-37827.

1096. Thompson, D. G. and Gunnala, S., "High temperature properties of solid tantalum chip capacitors", IEEE Trans. Parts, Hybrids & Packag., PHP-13 (4), 390-4 (1977), EEA81-9960.

1097. Tomago, A., Shimizu, T., Iijima, Y. and Yamauchi, I., "Development of oil-impregnated, all-polypropylene-film power capacitor", IEEE Trans. Electr. Insul., EI-12 (4), 293-300 (1977), EEA80-35791.

1098. Uemura, T. and Onoda, S., "Solid tantalum capacitors: small in size, but capable of high power", JEE, no. 119, 49-51 (1976), EEA80-20548.

1099. Veloric, H. S., Mitchell, J., Jr., Theriault, G. E. and Carr, L. A., Jr., "Capacitors for microwave applications", IEEE Trans. Parts, Hybrids & Packag., PHP-12 (2), 83-9 (1976).

1100. Vonkampen, T. and Reeves, P., "Consider polypropylene capacitors", Electron. Des., 25 (20), 76-7 (1977), EEA81-23526.
1101. Walker, J., "Electrolytic capacitors ... the design revolution continues", Electron. Equip. News, 13-15 (1977), EEA81-28069.
1102. Wallace, C. L. and Jones, R. B., "Process possibilities using thermoplastic ink vehicles in the fabrication of multilayer ceramic capacitors", NTG-Fachbr., no. 60, 26-7 (1977), German, EEA81-23634.
1103. Watase, S., "Aluminum electrolytic capacitors: explosion-proof and noninflammable", JEE, no. 119, 45-8 (1976), EEA80-20547.
1104. Webinger, R., "Transient load capability of aluminium electrolytic capacitors", Components Rep., 13 (1), 6-9 (1978), EEA81-45674.
1105. Werbizky, G. G., "Tantalum capacitors", IBM Tech. Disclosure Bull., 20 (5), 1725 (1977), EEA81-32939.
1106. Wessberg, W. N. and Ewell, G. J., "Special lot acceptance tests for multilayer ceramic chip capacitors", 27th Electronic Components Conference, 1977, p. 452-7, EEA80-35801.
1107. White, R., "Aluminium and plastic capacitors: an overview", Electron, no. 114, 46, 49 (1977), EEA80-35786.
1108. Wyatt, P. W., "DC field dependence of tantalum film capacitors", J. Electrochem. Soc., 123 (5), 667-75 (1976).
1109. Wyatt, P. W., "Aging of highly N-doped α-Ta thin-film capacitors", IEEE Trans. Components, Hybrids & Manuf. Technol., CHMT-1 (2), 148-51 (1978), EEA81-33019.
1110. Yamaoka, Y., "Ceramic capacitors advancing toward high quality and low cost", JEE, no. 119, 43-4 (1976), EEA80-20546.
1111. Yoneda, H. and Miyazaki, Y., "Aluminum electrolytic capacitors for high frequency applications", Natl. Tech. Rep. Matsushita Electr. Ind., 24 (1), 68-76 (1978).
1112. Yoshimura, S., Itoh, Y., Yasuda, M., Murakami, M., Takahasi, S. and Hasegawa, K., "Aluminum solid electrolytic capacitor with an organic semiconductor electrolyte", IEEE Trans. Parts, Hybrids & Packag., PHP-11 (4), 315-21 (1975).
1113. Young, P. L., "Thin film capacitor and method", Patent USA 4002542, Publ. January 1977.
1114. Young, P. L., "Method of forming a thin film capacitor", Patent USA 4038167, Publ. July 1977.
1115. Young, P. L., "Thin film tantalum oxide capacitor", Patent USA 4089039, Publ. May 1978.

XI.   EPOXIES

1116. Abshier, C. S., Berry, J. and Maget, H. J. R., "Toughening agents improve epoxy encapsulants", Insul./Circuits, 23 (11), 27-9 (1977), EEA81-13303.
1117. Anon., "Corrosion-resistant polymers surpass epoxy dielectrics", Des. Eng., 23 (1975), EEA79-13065.

1118. Anon., "Packaging and epoxy resins for packaging electrical
      and electronic components", Patent Japan 75-144753, Publ.
      November 1975, CA85-136065.
1119. Anon., "Using sealants, encapsulants, and adhesives", Elec-
      tron. Prod., 6 (4), 37, 39 (1977), EEA81-9246.
1120. Antonen, R. C., Michael, K. W. and Engelman, J. H., "New
      hybrid electronic molding compounds", J. Electron. Mater.,
      6 (1), 49-60 (1977), EEA80-35002.
1121. Beach, W. F., "Parylene and microelectronics", Electrochemical
      Society Spring Meeting, Extended Abstracts, 1975, p. 309-11,
      EEA79-93.
1122. Berkner, R., "Electrically insulating encapsulating composi-
      tion for semiconductor devices", Patent USA 3931026, Publ.
      January 1976, CA84-158845.
1123. Bolger, J. C. and Morano, S. L., "Epoxy encapsulation com-
      pounds with improved thermal shock and reversion resistance",
      12th Electrical/Electronics Insulation Conference, 1975,
      p. 266-9, EEA79-40778.
1124. Cadenhead, R. L., "Substrate attach epoxies", Solid State
      Technol., 18 (10), 53-5 (1975), EEA79-4726.
1125. Catsiff, E. H. and Seltzer, R., "Hydantoin epoxy resins",
      Mod. Plast., 55 (7), 54-6, 58 (1978).
1126. Darrow, R. E., Memis, I. and Poliak, R. M., "Flexible epoxy
      material for sealing integrated circuit packages", IBM Tech.
      Disclosure Bull., 20 (10), 3860 (1978).
1127. Day, M. R. and Newton, J. R., "Polyesters bid to increase
      packaging applications", Mod. Packag., 49 (2), 28, 30, 33
      (1976).
1128. Delmonte, J., "Urethane or epoxy? Which encapsulant to use",
      Insul./Circuits, 23 (11), 37-8 (1977), EEA81-13304.
1129. Dickson, D. D., Jr., "Silicone elastomers for electronic
      packaging", Natl. SAMPE Symp. Exhib., 21, 533-44 (1976),
      CA87-69444.
1130. Domininghaus, H., "Selection criteria for plastics. III.",
      Feinwerktech. & Messtech., 85 (4), 180-7 (1977), German,
      EEA81-393.
1131. Donaldson, P. E. K. and Sayer, E., "Silicone-rubber adhe-
      sives as encapsulants for microelectronic implants; effect of
      high electric fields and of tensile stress", Med. and Biol.
      Eng. and Comput., 15 (6), 712-15 (1977), EEA81-21401.
1132. Dusek, K., Plestil, J., Lednicky, F. and Lunak, S., "Are
      cured epoxy resins inhomogeneous?", Polymer, 19 (4), 393-7
      (1978).
1133. Elsby, T. W., "Thermal characterization of epoxy and alloy
      attachment of hybrid components", 27th Electronic Components
      Conference, 1977, p. 320-3, EEA80-35951.
1134. Fossey, D. J., Smith, C. H. and Wischmann, K. B., "New pot-
      ting material - expandable polystyrene bead foam", SPI, Cell
      Plast. Div., Int. Cell Plast. Conf., 4th and Annual Conf.,
      19th, 1977, p. 141-6.

1135. Fritzen, P. and Planting, P. J., "Composite epoxy glass-microsphere dielectric for hermetic SMA type RF connectors", Proc. of the 13th Electr./Electron. Insul. Conf., 1977, p. 49-55.

1136. Hawkins, J. W., "New liquid silicone rubber products for electronics applications", Proceedings of the 13th Electrical/Electronics Insulation Conference, 1977, p. 25-9, EEA81-14886.

1137. Howe, M. A., Jr. and Weinberg, A. S., "Key attributes of the more commonly available shrink packaging materials", Package Dev. Syst., 7 (3), 26-8 (1977).

1138. Ichinohe, S., "Analytical methods for plasticizers in packaging materials", Yukagaku, 25, 670-7 (1976), CA86-44237.

1139. Kalina, J. F., "In films for packaging, the future belongs to composite structures", Mod. Plast., 53 (3), 36-8 (1976).

1140. King, N. E. and Andrews, E. H., "Fracture energy of epoxy resins above $T_g$", J. Mater. Sci., 13 (6), 1291-1302 (1978).

1141. Kitamura, M., Matsuda, Y. and Suzuki, H., "Resin for packaging semiconductor devices", Patent Japan 76-02385, Publ. January 1976, CA85-39925.

1142. Komarova, L. I., Salazkin, S. N., Bulgakova, I. A., Malaniya, M. I., Vinogradova, S. V. and Korshak, V. V., "Interaction of polyesters with epoxy polymers: Insertion of oxirane rings into ester bonds", J. Polym. Sci. Polym. Chem. Ed., 16 (7), 1643-57 (1978).

1143. Kookootsedes, G. J., "Packaging materials, techniques: Keys to high device reliability", Electrochemical Society Spring Meeting, Extended Abstracts, 1975, p. 307-8, EEA79-92.

1144. Kume, N., Matsuo, S. and Sawada, T., "Application of resin cured by ultraviolet light to electronic components", Natl. Tech. Rep., 23 (2), 257-81 (1977), Japanese, EEA81-9252.

1145. Laffargue, C. and Lahaye, J., "Electrodeposition of epoxy and acrylic polymers", J. Coat Technol., 49 (634), 85-90 (1977).

1146. Lemeunier, P., "Plastics for electronics", Electron. & Microelectron. Ind., no. 223, 47-52 (1976), French, EEA79-33177.

1147. Littlewood, S. and Briggs, B. F. N., "Investigation of current-interruption by metal-filled epoxy resins", J. Phys. D, 11 (10), 1457-62 (1978), EEA81-42113.

1148. Ludeck, W., "Temperature resisting plastic materials for the electrical engineering", Feingeraetetechnik, 26 (8), 362-4 (1977), German, EEA81-392.

1149. Lunak, S., Dusek, K. and Vladyka, J., "Effect of diffusion control in the glass transition region on critical conversion at the gel point during curing of epoxy resins", Polymer, 19 (8), 931-3 (1978).

1150. McClain, R. R., "High-reactivity epoxy resins", Adhes. Age, 21 (2), 31-4 (1978).

1151. McKague, E. L., Jr., Reynolds, J. D. and Halkias, J. E., "Swelling and glass transition relations for epoxy matrix material in humid environments", J. Appl. Polym. Sci., 22 (6), 1643-54 (1978).

1152. Marshall, C., "Epoxies in thick film hybrid circuits", Circuit World, 3 (4), 25-8 (1977), EEA81-14321.
1153. Martin, D. J., "Filled epoxy resins on the right track", Electr. Rev., 202 (20), 22-3 (1978), EEA81-41651.
1154. Matsuura, I., Yamaguchi, F. and Iwama, K., "Low melting point compositions for sealing semiconductor devices", Patent Japan 76-33112, Publ. March 1976, CA85-86274.
1155. Matsuura, I., Yamaguchi, F. and Iwama, K., "Low melting point sealing compounds for semiconductor devices", Patent Japan 76-81814, Publ. July 1976, CA85-136075.
1156. Michael, K. W., Bank, H. M. and Antonen, R. C., "New hybrid electronic molding compounds", 26th Electronic Components Conference, 1976, p. 382-5, EEA79-37733.
1157. Mitchell, C. and Berg, H., "Use of conductive epoxies for die attach", Proc. Int. Microelectron. Symp., 1976, p. 52-8, CA86-172623.
1158. Mock, W., Jr. and Holt, W. H., "Shock-wave compression of an alumina-filled epoxy in the low gigapascal stress range", J. Appl. Phys., 49 (3), 1156-8 (1978).
1159. Muller, K., "Impregnating and casting agents in the field of electronics", Beck Isoliertech., 23 (50), 23-33 (1975), EEA79-16911.
1160. Nieberlein, V. A., "Thermal conductivity enhancement of epoxies by the use of fillers", IEEE Trans. Components, Hybrids & Manuf. Technol., CHMT-1 (2), 172-6 (1978), EEA81-32240.
1161. Palmisano, R. and Neily, D. W., "Particulate silicone rubber an effective, removable encapsulant for electronic packaging", Report AD-A031162/1SL, Harry Diamond Labs., Adelphi, Md., 1976, 17 pp., CA86-107792.
1162. Pelmore, J. M., "The ultrasonic properties of some filled epoxy materials", Ultrasonics International, 1977, p. 316-22, EEA81-32471.
1163. Rauhut, H. W., "Advanced epoxy molding compounds for semiconductor encapsulation", 34th SPE Annual Technical Conference, Proceedings, 1976, p. 313-16.
1164. Redemske, R. F., "Use of epoxies in hybrid microcircuits", Proceedings of a Workshop on Reliability Technology for Cardiac Pacemakers. II, 1976, p. 14-15, Publ. 1977, EEA81-33033.
1165. Reinhart, J., "Epoxy molding compounds for semiconductor encapsulation", Circuits Manuf., 17 (8), 29-31 (1977), EEA81-18361.
1166. Rhodes, M. S., "Polyanhydride flexibilizing hardeners for epoxy resins", Insul./Circuits, 23 (13), 39-41 (1977).
1167. Rifkin, A. A., Schiller, J. M. and Turetzky, M. N., "Sealant for semiconductor package", IBM Tech. Disclosure Bull., 20 (9), 3446 (1978).
1168. Roylance, D. and Roylance, M., "Weathering of fiber-reinforced epoxy composites", Polym. Eng. Sci., 18 (4), 249-54 (1978).
1169. Savla, M. and Skeist, I., "Epoxy resins", High Polym., 29, 582-641 (1977).

1170. Semikron Gesellschaft fur Gleichrichterbau und Elektronik m.b.H, "Encapsulating composition for semiconductors", Patent UK 1421149, Publ. January 1976, CA84-143816.

1171. Turetzky, M. N., "Organic inhibitors of indium corrosion", IBM Tech. Disclosure Bull., 20 (9), 3445 (1978).

1172. Ueyama, T., Ueda, T., Kitayama, T., Fujita, T., Koshizuka, K. and Tatara, N., "Semiconductor resin composition", Patent Japan 76-50495, Publ. May 1976, CA85-103060.

1173. Vynckier, S. A., "A polyester range called 'Atlas'", Belgelectro-Export, no. 34, 13-15 (1977), EEA81-391.

1174. Wehman, T. C. and Bates, W. F., "Epoxy encapsulants for semiconductors", Electrochemical Society Fall Meeting, Extended Abstracts, 1976, p. 421-3, EEA80-36210.

1175. Wischmann, K. B. and Assink, R. A., "Removable encapsulant – polystyrene foam", SAMPE J, 13 (12), 15-19 (1977).

1176. Wolfe, G., "Composite epoxy material: a substitute for FR-4/G-10", Circuits Manuf., 17 (5), 58-64 (1977), EEA81-13598.

XII.  SCREENS, PASTES, AND INKS

1177. Allen, J. L., "Microwave applications of thick-film technology", Proceedings of the 1976 IEEE Southeastern Region 3 Conference on Engineering in a Changing Economy, 1976, p. 219-20, EEA79-37651.

1178. Anon., "Thick film hybrid technology: low-cost high-power modules", Electron. & Microelectron. Ind., no. 212, 45 (1975), French, EEA79-13049.

1179. Anon., "Gold pastes for microelectronics packaging cavities", Patent Japan 75-00398, Publ. January 1975, CA84-188568.

1180. Anon., "Electrically conductive adhesives and coatings", Electron. Prod. Methods & Equip., 5 (4), 42, 45, 49 (1976), EEA79-31905.

1181. Bakewell, J. J., "Method of making a bimetal screen for thick film fabrication", Patent USA 4033831, Publ. July 1977.

1182. Baudry, H., "The rheological control of pastes to determine their suitability for serigraphical applications", Electron. & Microelectron. Ind., no. 226, 38-41 (1976), French, EEA80-6428.

1183. Baudry, H. and Monneraye, M., "Screen-printed copper conductors for hybrid circuitry", 1st European Solid State Circuits Conference – ESSCIRC, Extended Abstracts, 1975, p. 84, EEA79-28711.

1184. Beeler, J., "Screenprinting thick-film", Circuits Manuf., 15 (5), 32-3 (1975), EEA79-1027.

1185. Berdov, G. I. and Koganitskaya, E. V., "Paste for joining a metal to Al$_2$O$_3$ ceramics", Patent USSR 541825, Publ. January 1977.

1186. Bolon, D. A., Lucas, G. M. and Schroeter, S. H., "Radiation curable conductive ink", IEEE Trans. Electr. Insul., EI-13 (2), 116-21 (1978), EEA81-23573.

1187. Brady, J. J., Ferrante, J. A., Milkovich, S. A. and Urfer, E. N., "Screening mask for depositing large area paste deposits", IBM Tech. Disclosure Bull., 20 (9), 3429-30 (1978), EEA82-633.

1188. Brassell, G. W. and Fancher, D. R., "Electrically conductive epoxies for screen printing applications", 27th Electronic Components Conference, 1977, p. 226-31, EEA80-35921.

1189. Brewer, D. H., "Analysis of thick film conductor inks", Report BDX-613-1491, Bendix Corp., Kansas City, Mo., 1976, 23 pp.

1190. Burns, R. W. and Turnbaugh, J. E., "Transient high-power stability of thick film resistor pastes", International Micro-electronics Conference, Proceedings of the Technical Program, 1975, p. 135-45.

1191. Chuiko, G. M. and Bobrysheva, G. M., "Paste for metallization of ceramics", Patent USSR 536141, Publ. November 1976.

1192. Coronis, D. H., "Pre-sensitized emulsion screens for thick-film printing", Electron. Packag. & Prod., 18 (7), 129-36 (1978).

1193. Crandell, T. L. and Sergent, J. E., "Some aspects of the measurement of the adhesion of thick-film conductor inks to alumina substrates", Proceedings of the 1977 International Microelectronics Symposium, p. 16-19.

1194. Creter, P., "Analyzing thick film inks", Circuits Manuf., 16 (9), 56, 58, 60-65 (1976), EEA80-9197.

1195. Custer, W. D., "UV curable screen inks", Ind. Finish Surface Coatings, 28 (333), 5-9 (1976).

1196. Dahlstrom, M. S. and Simon, P., "Thin-film patterning tech-nique for ceramic parts", IBM Tech. Disclosure Bull., 19 (12), 4753 (1977), EEA81-10037.

1197. Davison, J. W., "Cost effective screen printing", Circuit World, 3 (1), 53-5 (1976), EEA80-32230.

1198. Desai, K. S., "Technique for controlled paste filled via sur-face for preventing paste transfer at lamination of MLC green-sheets", IBM Tech. Disclosure Bull., 20 (9), 3438 (1978).

1199. DiGiacomo, G., Gniewek, J. J. and Rizzuto, J. B., "Substrate electrode paste", IBM Tech. Disclosure Bull., 21 (1), 170 (1978).

1200. Dubey, G. C., "Masks for printing thick-film circuits", Microelectron. & Reliab., 16 (1), 69-73 (1976), EEA80-20582.

1201. E. I. Du Pont de Nemours & Co., "Vehicles for metalizing compositions", Patent UK 1419550, Publ. December 1975.

1202. Engelhard Minerals & Chemical Corp., "Pastes for forming electrically conductive films on nonconductive substrates", Patent UK 1496994, Publ. January 1978.

1203. Fefferman, G., "UV-curable solder masks for flexible cir-cuits", Circuits Manuf., 18 (2), 22, 24, 26, 28 (1978), EEA81-49490.

1204. Franconville, F., "Screens: essential tools for thick film printing", Electron. & Microelectron. Ind., no. 209, 44-9 (1975), EEA79-8988.

1205. Freudenheim, H., "Screen printing stencils for thick film circuits", Electron. Prod. Methods & Equip., 4 (7), 37-9 (1975), EEA79-4732.

1206. Gibson, D. M., "A method for thick-film printing of conductor fine lines and spacings", Proceedings of the 1977 International Microelectronics Symposium, p. 277-80.

1207. Goldman, O. H. and Scheinberg, S., "Control of paste spreading in screening processes", IBM Tech. Disclosure Bull., 19 (11), 4263-4 (1977), EEA81-14276.

1208. Graves, J. F., "Thick film conductor materials - production qualification requirements and test procedures", Proceedings of the 1977 International Microelectronics Symposium, p. 155-61.

1209. Greenstein, B., "Vehicle and printing pastes for use in the manufacture of microelectronic packages", Patent USA 3975201, Publ. August 1976, CA85-136083.

1210. Greenstein, B., "Printing paste vehicle, gold dispensing paste and method of using the paste in the manufacture of microelectronic circuitry components", Patent USA 4032350, Publ. June 1977, CA87-61594.

1211. Grier, J. D., "A case for copper thick-film paste", Electron. Packag. & Prod., 17 (6), 58-61 (1977), EEA81-14339.

1212. Harris, H. R., "Comparison of thick-film screen materials", Electron. Packag. Prod., 16 (1), 60-2, 64 (1976).

1213. Henderson, J. T., "IC screening, reliability of ripoff?", Proceedings 1976 Annual Reliability and Maintainability Symposium, p. 452-5, EEA79-41575.

1214. Hilson, D. G. and Johnson, G. W., "New materials for low cost thick film circuits", Solid State Technol., 20 (10), 49-54 (1977).

1215. Hitch, T. T., "Gold- and silver-based thick-film conductors", 27th Electronic Components Conference, 1977, p. 260-8, EEA80-35923.

1216. Hitch, T. T., Whitaker, H. H., Botnick, E. M. and Goydish, B. L., "Chemical analyses of thick-film gold conductor inks", IEEE Trans. Parts, Hybrids & Packag., PHP-11 (4), 248-53 (1975), EEA79-13004.

1217. Hoffman, L. C., McMunn, C. W., Mones, A. H. et al, "Gold conductor compositions", Patent USA 4004057, Publ. January 1977.

1218. Imhof, M. and Kersuzan, M., "Non-precious metals for replacing gold in thick film inks", Electron. & Microelectron. Ind., no. 212, 46-7 (1975), French, EEA79-13001.

1219. Kaplan, P., Kelly, J. E. and Miller, L. F., "Thick film conductive paste system", IBM Tech. Disclosure Bull., 21 (5), 1864 (1978).

1220. King, R. P. and Nichols, C., "Screening masks and method of fabrication", IBM Tech. Disclosure Bull., 20 (2), 577-8 (1977), EEA81-18353.

1221. Kocsis, A., Leroux, A., Richard, S. and Aube, G., "Technolog-
      ical requirements for portable thick film hybrid devices",
      12th Electrical/Electronics Insulation Conference, 1975,
      p. 36-9, EEA79-37739.
1222. Krasovskaya, A. K. and Kilmshina, I. M., "Paste for metalli-
      zation of $Al_2O_3$ ceramic", Patent USSR 481580, Publ. August
      1975.
1223. Kuist, C. H., "Conductive elastomers solve interconnection
      problems", Electron. Packag. Prod., 16 (2), 81-2, 84 (1976).
1224. Kuskovskaya, I. I., Antoshina, A. S. and Chelnokov, E. I.,
      "Paste containing W and $Al_2O_3$ for metallization of high-
      alumina ceramic", Patent USSR 537989, Publ. December 1976.
1225. Kuzel, R. and Hoschl, P., "Microscopic structure of cermet
      layers", Jenno Mech. & Opt., 21 (11), 333-5 (1976), Czech,
      EEA80-24618.
1226. Lemeunier, P., "Copper pastes used for thick film hybrids",
      Electron. & Appl. Ind., no. 238, 41-3 (1977), French,
      EEA81-10048.
1227. Lemon, T. H., "High-tensile-strength thick-film silver-
      palladium metallizations - improved compositions for screen-
      printed circuits", Platinum Met. Rev., 19 (4), 46-53 (1975).
1228. Leven, S. S., "Screening procedure for adhesion degradation
      due to solder leaching in thick-film hybrid microcircuits",
      Solid State Technol., 20 (3), 39-44 (1977), EEA80-24625.
1229. Lucas, J., "Inks for thick printed circuit fabrication",
      Patent Japan 74-58903, Publ. June 1974, CA84-158878.
1230. Maxwell, W. E. and Hamill, W. A., "Ink selection for elec-
      tronic product printing", Electron. Packag. & Prod., 17 (7),
      93-4, 96, 98, 101 (1977).
1231. Miller, L. F., "2, 2, 4 trimethyl pentanediol 1, 3 mono-
      isobutyrate in grit pastes", IBM Tech. Disclosure Bull., 20
      (8), 3070 (1978).
1232. Miller, L. F., "E.H.E.C. in MLC pastes", IBM Tech. Dis-
      closure Bull., 20 (8), 3071 (1978).
1233. Nordstrom, T. V. and Yost, F. G., "Sintering behaviour of a
      reactively bonded thick film gold ink", J. Electron. Mater.,
      7 (1), 109-22 (1978), EEA81-37914.
1234. Onufer, R. J. and Stanley, E. C., "Using a rotational visco-
      meter to characterize thick film materials", Insul./Circuits,
      22 (1), 23-7 (1976).
1235. Owens-Illinois, Inc., "Gold pastes for microelectronics
      packaging cavities", Patent Japan 75-00398, Publ. January
      1975, CA84-188568.
1236. Philips Electronic & Associated Industries, Ltd., "Mask for
      silk-screen printing", Patent UK 1374468, Publ. November 1974.
1237. Philips Electronic & Associated Industries, Ltd., "Silk screen
      printing paste", Patent UK 1489031, Publ. October 1977.
1238. Polinsky, P. W., "High-speed thick film printing: assessing
      the critical variables", Circuits Manuf., 17 (10), 26-30 (1977).

1239. Rizzuto, J. B. and Young, S. P., "Metallizing paste", IBM Tech. Disclosure Bull., 21 (2), 553 (1978).

1240. Robertson, S. D., Licari, J. J. and Buckelew, R. L., "Marginal wire bonds and the effectiveness of MIL-STD-883 screens", Proceedings of the 1977 International Microelectronics Symposium, p. 146-9.

1241. Schuster, R., "Establishing an in-house screening lab", Eval. Eng., 15 (2), 14-6 (1976), EEA79-27518.

1242. Shah, J. S. and Hahn, W. C., "Material characterization of thick-film resistor pastes", IEEE Trans. Components, Hybrids & Manuf. Technol., CHMT-1 (4), 383-92 (1978).

1243. Siemens, A. G., "Thick-film circuits", Patent UK 1422085, Publ. January 1976.

1244. Stroms, K. F., "Planographic paste application to sheet type parts", IBM Tech. Disclosure Bull., 20 (11), 4360-1 (1978).

1245. Suppelsa, A. and Khadpe, S., "Performance characteristics of platinum-silver conductor materials in hybrid microcircuits", 26th Electronic Components Conference, 1976, p. 150-5, EEA79-37723.

1246. Tobita, T., Nakayama, T. and Takasago, H., "Application of silver-palladium conductor paste and solder paste to thick hybrid IC's", Denshi Tsushin Gakkai Gijutsu Kenkyu Hokoku, 676 (6), 47-55 (1976), Japanese, CA86-36738.

1247. United Kingdom Atomic Energy Authority, "Metalizing pastes", Patent UK 1378520, Publ. December 1974.

1248. Usami, T., "Electroconductive paste for forming printed circuits on ceramic printed circuit boards", Patent Japan 75-134197, Publ. October 1975, CA84-115088.

1249. Vitriol, W. A. and Hodge, P. M., "Sophisticated techniques solve ink manufacturing problems", Solid State Technol., 19 (3), 33-7 (1976), EEA79-28646.

1250. Weitze, A. and Leskovar, P., "Thick-layer conductor path pastes", Patent USA 4039721, Publ. August 1977.

1251. Yarmolinskaya, L. N., Lyzlova, T. N. and Levin, A. S., "Molybdenum-based paste for metallization of ceramics", Patent USSR 529143, Publ. September 1976.

XIII. COATING/PASSIVATION

1252. Amaro, J. M., Hultmark, E. B., Natoli, T. N. and Puttlitz, K. J., "Back sealing electronic modules", IBM Tech. Disclosure Bull., 18 (3), 751-2 (1975), EEA79-85.

1253. Anon., "Protective coatings for hybrids, resistors and auto electronics", Circuits Manuf., 15 (8), 16, 18, 20 (1975), EEA79-7831.

1254. Anon., "Vacuum coatings combat corrosion", Des. Eng., 45, 47, 49 (1977), EEA80-34983.

1255. Anon., "Passivation coatings for thin film resistors", Circuits Manuf., 17 (8), 16, 18, 22, 24 (1977), EEA81-19152.

1256. Antonucci, R. F., Lahey, E. and Schiller, J. M., "Substrate protective coating process", IBM Tech. Disclosure Bull., 21 (6), 2429 (1978).

1257. Asada, E. and Kaneko, K., "Coating compositions containing ruthenium oxide", Patent Japan 77-52933, Publ. April 1977, CA87-119431.

1258. Benninghoff, H., "Coated materials - Sn, Sn-Pb, and Pb coatings for functional applications in electrotechnics and electronics", Tech. Rundsch., 69 (20), 13, 15 (1977), German, EEA80-32231.

1259. Blust, H. L., Lindburg, N. L. and Henry, D. V., "Method of anchoring metallic coated leads to ceramic bodies and lead-ceramic bodies formed thereby", Patent USA 3970235, Publ. July 1976.

1260. Boeckl, R. S., "Spin coating process for prevention of edge buildup", Patent USA 4068019, Publ. January 1978.

1261. Busch, R. and Bayne, M. A., "Development of sputtered high temperature coatings for thrust chambers", Report N77-12181/2SL, Battelle Pacific Northwest Lab., Richland, Wash., 1976, 25 pp.

1262. Butler, J. F. and Babyak, W. J., "Method and apparatus for vapor-depositing coatings on substrates", Patent USA 3989862, Publ. November 1976.

1263. Carlson, D. E. and Goodman, L. A., "Method for forming electrode patterns in transparent conductive coatings on glass substrates", Patent USA 3991227, Publ. November 1976.

1264. Cherezova, L. A. and Kryzhanovskii, B. P., "Microcomposite coatings based on nitrides of refractory metals", Inorg. Mater., 13 (1), 147-8 (1977), EEA81-852.

1265. Comizzoli, R. B., "Corona discharge - electrostatic method for deposition of powdered passivation glass on semiconductor devices", IEEE Trans. Parts, Hybrids & Packag., PHP-13 (3), 322-8 (1977).

1266. Comizzoli, R. B., "Nondestructive, reverse decoration of defects in IC passivation overcoat", J. Electrochem. Soc., 124 (7), 1087-95 (1977), EEA80-41526.

1267. Dargavel, G., "Packaging electronic circuits. I. Casting and coating", Electron, no. 103, 49-50 (1976), EEA80-40.

1268. DiGiacomo, G., "Passivation of lead/indium solder", IBM Tech. Disclosure Bull., 20 (6), 2314 (1977).

1269. Dittrich, F. J., Smyth, R. T. and Weir, J. D., "Production of electronic coatings by thermal spraying", Proceedings of the 1977 International Microelectronics Symposium, p. 274-6.

1270. Funakawa, S. and Yamane, M., "Glass passivated junction semiconductor devices", Patent USA 4080621, Publ. March 1978.

1271. Goryachkin, V. A. and Florianovich, G. M., "Effect of bichromate and molybdate ions on the passivation potential of chromium", Prot. Met., 13 (6), 590-2 (1977).

1272. Goulding, T. and Orton, L., "Coating composition for packaging films", Patent UK 1468675, Publ. March 1977, CA87-119429.

1273. Grisik, J. J., "Diffusion coating method", Patent 4004047, **Publ. January 1977.**

1274. Heap, B. C. and France, S. A., "Improved hybrid circuit assembly yields and reliability by glassivation of the semiconductor chip", Electrocompon. Sci. & Technol., 4 (3-4), 117-24 (1977), EEA81-14334.

1275. Heid, K., "Nichrome passivation coatings - help or hindrance?", 27th Electronic Components Conference, 1977, p. 68-71, EEA80-35909.

1276. Hieber, K. and Stolz, M., "Coating the inside surfaces of hollow sections with a tantalum layer for corrosion protection", Siemens Forsch.- & Entwicklungsber., 6 (4), 232-5 (1977), EEA80-34991.

1277. Hill, L. W., "Stress analysis - a tool for understanding coatings performance", Prog. Org. Coatings, 5 (3), 277-94 (1978).

1278. Hoffman, H. S., "MC cermet $SiO_2$ passivation process", IBM Tech. Disclosure Bull., 20 (7), 2630 (1977), EEA81-45672.

1279. Hutson, J. L., "Technique for passivating semiconductor devices", Patent USA 4007476, Publ. February 1977, CA86-114494.

1280. International Computers, Ltd., "Glass coating, bonding, or encapsulation", Patent UK 1369458, Publ. October 1974.

1281. Iwasa, M. and Horikawa, K., "Electrostatic powder coating", Patent Japan 77-52945, Publ. April 1977, CA87-119432.

1282. Kale, V. S. and Riley, T. J., "A production parylene coating process for hybrid microcircuits", IEEE Trans. Parts, Hybrids & Packag., PHP-13 (3), 273-9 (1977), EEA81-911.

1283. Kato, H., Yagi, H. and Fukuta, S., "Method for coating a conductive material", Patent USA 3988231, Publ. October 1976.

1284. Kinmonth, R. A., Jr. and Norton, J. E., "Effect of spectral energy distribution on degradation of organic coatings", J. Coat. Technol., 49 (633), 37-44 (1977).

1285. Losure, J. A., "Coating of vias in alumina substrates", Res./ Dev., 27 (8), 57, 59-60, 62 (1976), EEA79-40782.

1286. Losure, J. A., "Production coating of vias in alumina substrates with vacuum evaporated chromium and gold", 26th Elecrronic Components Conference, 1976, p. 1-8, EEA79-37717.

1287. Lublin, P. and Koffman, D., "Coating thickness monitor for multiple layers", Patent USA 3984679, Publ. October 1976.

1288. McClocklin, R. S. and Teal, B. A., "Thermal-spray coatings for computer components", J. Vac. Sci. & Technol., 12 (4), 784-5 (1975), CCA11-6395.

1289. Malinovski, Y. P., Bakardzhiev, S. T. and Martinov, G. M., "Vapor deposition apparatus for coating continuously moving substrates with layers of volatizable solid substances", Patent USA 4094269, Publ. June 1978.

1290. Mesrobian, R. B., "Ultraviolet applications for packaging", Int. J. Radiat. Phys. Chem., 9, 307-24 (1977), CA87-7504.

1291. Miller, S. C., "A technique for passivating thin film hybrids", Proc. Int. Microelectron. Symp., 1976, p. 39-44, CA86-198812.

1292. Morton, D. E., "Method of optical thin film coating", Patent
      USA 4058638, Publ. November 1977.
1293. Nickols, S. E. and Olsen, C. E., "Glass-metal-glass selectively
      passivated charge plate", IBM Tech. Disclosure Bull., 19
      (12), 4754 (1977), EEA81-9238.
1294. Okubo, H., Kato, M., Yoshida, M., Ito, A. and Kaetsu, I.,
      "New coating materials prepared by radiation-induced poly-
      merization - 1. Mar-resistant coating composition and prop-
      erties", J. Appl. Polym. Sci., 22 (2), 487-96 (1978).
1295. Paal, G. and Schackert, K., "Method of passivation and planar-
      izing a metallization pattern", Patent USA 4089766, Publ.
      May 1978.
1296. Peekema, R. M., "Charge plate passivation process", IBM Tech.
      Disclosure Bull., 20 (11), 2923 (1978).
1297. RCA Corp., "Method of selectively depositing glass on semi-
      conductor devices", Patent UK 1464682, Publ. February 1977.
1298. Rehfeld, L., "Vacuum deposited coatings", Patent UK 1420492,
      Publ. January 1976.
1299. Rie, J. E., Garstang, C. W. and Morgan, C. R., "Ultraviolet
      curable covercoat material for flexible circuitry", Proceed-
      ings of the 13th Electrical/Electronics Insulation Conference,
      1977, p. 15-19, EEA81-14289.
1300. Robertson, G. W. E., "Passivation and corrosion", Conference
      on the Operation of Instruments in Adverse Environments,
      1976, p. 19-25, Publ. 1977, EEA81-387.
1301. Segui, Y. B. A., "Microelectronic applications of plasma-
      polymerized films", Thin Solid Films, 50, 321-4 (1978),
      EEA81-41654.
1302. Subramanian, R. V., Jakubowski, J. J. and Williams, F. D.,
      "Interfacial aspects of polymer coating by electropolymeri-
      zation", J. Adhes., 9 (3), 185-95 (1978).
1303. Terada, T., Ushigome, M. and Tamaki, S., "Tantalum nitride
      thin film circuits on polymide", Electrochemical Society
      Spring Meeting, Extended Abstracts, 1975, p. 122-3, EEA79-1030.
1304. Tomozawa, A., Nakata, K., Kikuchi, A. and Agatsuma, T.,
      "Forming insulating film on interconnection layer", Patent
      USA 3935083, Publ. January 1976.
1305. Washo, B. D., "Rheology and modeling of the spin coating
      process", IBM J. Res. & Dev., 21 (2), 190-8 (1977),
      EEA81-13290.
1306. Whittington, J. K., Malloy, G. T., Mastro, A. R. and Hutch-
      ens, R. D., "Internal conformal coatings for microcircuits",
      IEEE Trans. Components, Hybrids & Manuf. Technol., CHMT-1
      (4), 416-22 (1978).
1307. Yerman, A. J., "Semiconductor element having a polymeric
      protective coating and glass coating overlay", Patent USA
      4017340, Publ. April 1977, CA86-198874.

XIV.  PLATING

1308. Acitelli, M. A., Alpaugh, W. A. and Woods, J. J., "Surface preparation for additive plating", IBM Tech. Disclosure Bull., 20 (9), 3392 (1978), EEA82-619.
1309. Ahn, K. Y. and Powers, J. V., "Plating base for nickel iron in bubble structures", IBM Tech. Disclosure Bull., 20 (9), 3784 (1978).
1310. Alpaugh, W. A. and McCreary, J. M., "Copper plating advanced multilayer boards", Insul./Circuits, 24 (3), 27-32 (1978), EEA81-33001.
1311. Anis, N., "ABCs of electroplating ABS", Plast. Eng., 33 (1), 14-17 (1977).
1312. Anon., "Plating of printed edgeboard contacts", Report IVF-Resultat-75641, Inst. Verkstadsteknisk Forskning, Gothenburg, Sweden, 1975, 11 pp., EEA81-23601.
1313. Anon., "Ammonia-free baths for bright and ductile plating", Circuits Manuf., 17 (5), 68-70 (1977), EEA81-10010.
1314. Barnes, C. and Wards, J. J. B., "The use of ultrasonic agitation in gold plating for electronic applications", Trans. Inst. Met. Finish., 55 (pt. 3), 101-3 (1977), EEA81-41304.
1315. Beeman, L., "Controlling hidden costs in quality plating-on-plastics", Ind. Finish, 53 (4), 46-50 (1977).
1316. Brooks, P., Farr, J. P. G., Sheppard, K. G. and Taylor D., "Silver plating in the manufacture of semi-conductor devices", Trans. Inst. Met. Finish, 56 (pt. 2), 75-80 (1978).
1317. Brookshire, R. L., "Advancement in brush plating", Electro-chemical Society Fall Meeting, Extended Abstracts, 1976, p. 711-12, EEA80-34997.
1318. Canestaro, M. J., "Additive plating coverage", IBM Tech. Disclosure Bull., 20 (9), 3391 (1978), EEA82-618.
1319. Caricchio, J. J., Jr. and York, E. R., "Pd-Cu plating system", IBM Tech. Disclosure Bull., 20 (5), 1721 (1977), EEA81-37130.
1320. Caricchio, J. J. and York, E. R., "Method and composition for plating palladium", Patent USA 4076599, Publ. February 1978.
1321. Caricchio, J. J. and York, E. R., "Method for plating palladium nickel alloy", Patent USA 4100039, Publ. July 1978.
1322. Carpenter, M. R. and Greene, K. F., "Gold barrel plating process", IBM Tech. Disclosure Bull., 21 (7), 2933-4 (1978).
1323. Castillero, A. W., "Gold plating alternatives [for printed circuits]", Electron. Packag. & Prod., 17 (4), 160-1 (1977), EEA81-5608.
1324. Chiriac, H. and Pop, G., "Annealing influence on the rotattional hysteresis losses of nickel electroplated films", Phys. Status Solidi (A) Appl. Res., 31 (1), K73-K76 (1975).
1325. Cohen, R. L., Raub, C. J. and Muramaki, T., "Investigation of agitation effects on electroplated copper in multilayer board plated-through holes in a forced-flow plating cell", J. Electrochem. Soc., 125 (1), 36-43 (1978).

1326. Costello, B. J., "Notes on fusing: the effect of thin-lead plating", Electron. Packag. & Prod., 18 (7), 64-6, 68, 70-1 (1978).

1327. Dafter, R. V., Jr., Haynes, R. and Rehrig, D. L., "High-speed, selective gold-plating of lead frames for microelectronic packaging. II. An evaluation of solutions for spot-plating of gold, using either direct current or asymmetrical AC waveforms", Western Electric Eng., 22 (2), 57-60 (1978), EEA81-48601.

1328. DiBari, G. A., "Plating on aluminum: Pretreatments and corrosion performance", Plat. Surf. Finish, 64 (5), 68-70, 72-4 (1977).

1329. Eickelberg, F., "New appeal of electroplating: costs are down, performance is up", Mod. Plast., 53 (12), 45-7 (1976).

1330. Eidschun, C. D., "How controlled solution plating saves gold", Ind. Finish., 54 (6), 36-9 (1978).

1331. Feldstein, N., "Method of electroless plating", Patent USA 3900599, Publ. August 1975.

1332. Flaskerud, P. and Mann, R., "Silver plating as a substitute system for gold in IC packaging", International Microelectronics Conference, Proceedings of the Technical Program, 1975, p. 104-21.

1333. Frankenthaler, J. J., Hastings, G. H., Jr. and Summa, W. J., "Additively plating on both sides of transparent laminates", IBM Tech. Disclosure Bull., 20 (9), 3393 (1978), EEA82-620.

1334. Fukutomi, M., Kitajima, M., Okada, M. and Watanabe, R., "Silicon nitride coating on molybdenum by RF reactive ion plating", J. Electrochem. Soc., 124 (9), 1420-4 (1977).

1335. Galyon, G. T., Lee, A. P., Martin, H. J. and Williams, T. O., "Simultaneous co-evaporation of thick metal films", IBM Tech. Disclosure Bull., 20 (9), 3448 (1978).

1336. Hadad, M. M., "Selectively electroplating pad terminals on an MLC substrate", IBM Tech. Disclosure Bull., 20 (9), 3443-4 (1978), EEA82-635.

1337. Haddad, M. M., "Immersion gold plating bath", IBM Tech. Disclosure Bull., 20 (11), 4768 (1978).

1338. Haddad, M. M. and Suierveld, J., "Depositing crack-free heavy electroless nickel coating", IBM Tech. Disclosure Bull., 20 (8), 3079 (1978).

1339. Hajicek, D. J., "Quantitative determination of plating thickness by scanning electron microprobe", Proceedings of the 8th Annual Meeting of the International Metallographic Society, 1975, Publ. in Microstructure Science, Vol. 4, 1976, p. 61-72.

1340. Hinoul, M., "The deposition of thick ruthenium layers by ion plating", Thin Solid Films, 45 (3), 539 (1977), EEA81-4996.

1341. Hulme, J. and Jordan, N. H., "Plating on plastics today - 1, 2", Prod. Finish, 29 (4), 8-10; ibid 29 (5), 55-8 (1976).

1342. Hulme, J. and Jordan, N. H., "Plating on plastics today", Finish Ind., 1 (7), 15-16 (1977).

1343. Kremer, R. E. and Ogle, K. L., "How to control nickel and gold plating thickness", Circuits Manuf., 17 (2), 28, 30, 32 (1977), EEA81-5604.

1344. Litsch, H. J. and Rehrig, D. L., "Asymmetric plating current", Tech. Dig., no. 43, 27 (1976), EEA80-9205.

1345. Loehndorf, D., "Electroplating process for forming heavy gold plating on multilayer ceramic substrates", IBM Tech. Disclosure Bull., 20 (5), 1740 (1977), EEA81-33008.

1346. Lynch, T. E. and York, E. R., "Non-cyanide gold nickel alloy plating", IBM Tech. Disclosure Bull., 21 (3), 947 (1978).

1347. Martinson, C. W. B., Nordlander, P. J. and Karlsson, S. E., "AES investigation of the interface between substrate and chromium films prepared by evaporation and ion plating", Vacuum, 27 (3), 119-23 (1977), EEA81-381.

1348. Mason, D. R. and Blair, A., "Comparison of cyanide and sulphite gold plating processes", Trans. Inst. Met. Finish., 55 (pt. 4), 141-8 (1978).

1349. Messner, G., "Imaging processes for electrolessly deposited conductors", Electrochemical Society Fall Meeting, Extended Abstracts, 1976, p. 700-1, EEA80-35886.

1350. Missel, L., "Safe, tolerant strike for plating on aluminum", Plat. Surf. Finish., 64 (7), 32-5 (1977).

1351. Missel, L. and Phipps, P. B., "Plating bath for nickel iron chromium", IBM Tech. Disclosure Bull., 21 (6), 2494-5 (1978).

1352. Mittal, K. L., "Deposition of metal on polymer substrate", IBM Tech. Disclosure Bull., 20 (5), 1824 (1977).

1353. Mouhot, A., "Device for studying the plating of printed circuit holes", IBM Tech. Disclosure Bull., 21 (3), 1076-7 (1978).

1354. Murayama, Y., "Thin film formation of $In_2O_3$, TiN and TaN by RF reactive ion plating", J. Vac. Sci. & Technol., 12 (4), 818-20 (1975), EEA79-8992.

1355. Murayama, Y. and Takao, T., "Structure of a silicon carbide film synthesized by R.F. reactive ion plating", Thin Solid Films, 40 (1-2-3), 309-17 (1977).

1356. Ohmae, N., Nakai, T. and Tsukizoe, T., "Ion-plated thin films for anti-wear application", International Conference on Wear of Materials, 1977, p. 350-7.

1357. Okinaka, Y., Turner, D. R., Wolowodiuk, C. and Graham, D. W., "Automatic electroless copper bath analyzer and controller using polarography", Electrochemical Society Fall Meeting, Extended Abstracts, 1976, p. 704-5, EEA80-35888.

1358. Oliver, G. C., "Plating fine lines with a nozzle [printed circuits]", Insul./Circuits, 24 (8), 23-4 (1978), EEA82-622.

1359. Patel, H. N., "Plating and tinning technique for screen AND/OR sputter rhodium conductor pads", IBM Tech. Disclosure Bull., 21 (7), 2987 (1978).

1360. Pcihoda, W. A. and Oberholtzer, B. D., "Automating a small-volume thin-film plating process", Prod. Finish, 41 (8), 54-7 (1977).

1361. Pcihoda, W. W. and Walker, A. E., Jr., "Production plating of thin-film circuits", Electron. Packag. & Prod., 15 (6), 74-6, 78 (1975).

1362. Poa, S. P., Wann, C. C. and Wu, C. J., "Study of the etching effect on the metal-to-ABS surface adhesion in electroless plating", Met. Finish., 75 (8), 13-16 (1977).

1363. Pridans, J. V. and Sekora, C. M., "Method for through-hole plating void elimination", IBM Tech. Disclosure Bull., 21 (6), 2267 (1978).

1364. Pushpavanam, M., "Electroless nickel: a versatile coating", Finish. Ind., 1 (6), 48 (1977).

1365. Pushpavanam, M. and Shenoi, B. A., "Electroless copper plating. I.", Finish. Ind., 1 (10), 36, 40, 43 (1977), EEA81-4774.

1366. Pushpavanam, M. and Shenoi, B. A., "Electroless copper plating. II.", Finish. Ind., 1 (11), 26, 28 (1977), EEA81-4775.

1367. Rapson, W. S. and Groenewald, T., "Use of gold in autocatalytic plating processes", Gold Bull., 8 (4), 119-26 (1975).

1368. Rehrig, D. L., "High-speed, selective gold-plating of lead frames for microelectronic packaging. I. Process development and evaluation of process parameters", Western Electric Eng., 22 (2), 48-56 (1978), EEA81-48600.

1369. Rothschild, B. F., "Effect of orthophosphate in copper pyrophosphate plating solutions and deposits", Met. Finish., 76 (1), 49-51 (1978).

1370. Rubinstein, M., "High speed selective brush plating - an overview", Electrochemical Society Fall Meeting, Extended Abstracts, 1976, p. 709-10, EEA80-34996.

1371. Rubenstein, M., "Precision circuit repair with selective plating", Insul. Circuits, 23 (8), 37-40 (1977).

1372. Sard, R., "Significant properties of selectively plated deposits and their measurement",, Electrochemical Society Fall Meeting, Extended Abstracts, 1976, p. 691, EEA80-34993.

1373. Schaer, G. R. and Safranek, W. H., "Selective high-speed gold plating", Electrochemical Society Fall Meeting, Extended Abstracts, 1976, p. 719-21, EEA80-35889.

1374. Sorensen, G. and Whitton, J. L., "Plating with ion accelerators", Vacuum, 27 (3), 155-7 (1977), EEA81-384.

1375. Spalvins, T., "New applications of sputtering and ion plating", ASME Paper No. 77-DE-21, 1977, 9 pp.

1376. Spiliotis, N. J., "Tin-lead plating practices that assure proper wetting upon solder reflow", Insul./Circuits, 24 (6), 33-6 (1978), EEA81-48748.

1377. Srivastava, R. D. and Mukerjee, R. C., "Electrodeposition of binary alloys: an account of recent developments", J. Appl. Electrochem., 6 (4), 321-31 (1976).

1378. Stokes, S., "An activity, chemistry and stress monitoring control system for electroless copper plating", Electrochemical Society Fall Meeting, Extended Abstracts, 1976, p. 706-7, EEA80-34994.

1379. Takao, M. and Tasaki, A., "Ferromagnetic thin films prepared by ion plating", IEEE Trans. Magn., MAG-12 (6), 782-4 (1976).

1380. Teer, D. G., "Adhesion of ion plated films and energies of deposition", J. Adhes., 8 (4), 289-300 (1977).

1381. Tench, D. and Ogden, C., "On the functioning and malfunctioning of dimercaptothiadiazoles as leveling agents in circuit board plating from copper pyrophosphate baths", J. Electrochem. Soc., 125 (8), 1218-24 (1978), EEA82-624.

1382. Turner, D. R., "Selective plating - an overview", Electrochemical Society Fall Meeting, Extended Abstracts, 1976, p. 680-1, EEA80-34992.

1383. Tuxford, A. M., "Plating base process for nickel iron plated bubble devices", IBM Tech. Disclosure Bull., 20 (9), 3782-3 (1978).

1384. van Nie, A. G., "Electroless NiP processing for hybrid integrated circuits", Microelectron. & Reliab., 15 (3), 221-6 (1976).

1385. White, M. S., "Mechanical plating", Prod. Finish., 30 (10), 9-10, 14 (1977).

1386. Wiesner, H. J., Comfort, W. J., III, Frey, W. P., Gardiner, C. and Kelley, W., "Electroless nickel plating of a large turbine rotor", Electrochemical Society Fall Meeting, Extended Abstracts, 1976, p. 708, EEA80-34995.

1387. Williams, E., "Ion plating - coat of many colors", New Sci., 74 (1055), 588-9 (1977), EEA80-31767.

1388. Williams, E. W., "Gold ion plating", Gold Bull., 11 (2), 30-4 (1978).

1389. York, E. R., "Palladium-gold plating treatment", IBM Tech. Disclosure Bull., 20 (2), 550 (1977), EEA81-18352.

XV.   ELECTRON BEAM TECHNIQUES

1390. Basalaeva, M. A. and Bashenko, V. V., "Movement of metal in the weld pool in electron-beam welding", Weld. Prod., 24 (3), 1-3 (1977).

1391. Bashkatov, A. V., "Effect of the composition of the joint on its geometrical characteristics in electron-beam welding", Weld. Prod., 24 (7), 38-40 (1977).

1392. Blaha, A., "Micromilling of thin films by an electron beam", Slaboproudy Obz., 39 (9), 408-16 (1978), Czech, EEA82-641.

1393. Di Mino, A., "Electronic trimming of microelectronic resistors", Patent USA 3617684, Publ. November 1971.

1394. Engel, J. M. and Holmstrom, F. E., "Electron beam testing", Quality Progress, 3 (11), 38-40 (1970).

1395. Hegner, F. and Feuerstein, A., "Aluminum-silicon metallization by rate controlled dual EB-gun evaporation", Solid State Technol., 21 (11), 49-54 (1978).

1396. Hughes, J. L., "Scale-up problems in e-beam evaporation and sputtering", J. Vac. Sci. & Technol., 15 (2), 305 (1978), EEA81-41636.

1397. Kashar, L., "Failure analysis of EB welded metal hybrid pack-
      ages", Proc. - Adv. Tech. Failure Anal. Symp., 1976, p. 80-5,
      CA86-11205.
1398. Konig, D., "New results in the field of micro-machining by
      means of electron beams", Vakuum-Tech., 16 (3), 48-52 (1967),
      German, EEA70-14865.
1399. Konov, A. N., Kolpakov, A. I. and Rafaevich, B. D., "Unit for
      electron-beam cutting of semiconductor plates", Instrum. &
      Exp. Tech., 20 (1), 293-5 (1977), EEA81-14617.
1400. Koste, W. W., "Electron beam processing on interconnection
      structures in multilayer ceramic modules", Met. Trans., 2
      (3), 729-31 (1971).
1401. Landig, T., McKoon, R. and Young, M., "Electron-beam melting
      of Ti-6Al-4V", J. Vac. Sci. & Technol., 14 (3), 808-14 (1977).
1402. Marin, G. and Simionescu, C., "Electron-beam vacuum brazing
      in glass to metal joints production", Stud. Cercet. Fiz., 24,
      137 (1972), Rumanian, EEA75-23434.
1403. Menesi, J., "Electron beam welding of metal and ceramics",
      Finommechanika, 12, 12 (1973), Hungarian, EEA76-19007.
1404. Muller, R. S., "Electron beam probing of semiconductor mater-
      ials and devices", Symposium on Test Methods and Measurements
      of Semiconductor Devices, Vol. 2, Budapest, April 25-28,
      (1967).
1405. Munakata, C., "An electron beam method of measuring diffusion
      voltage in semiconductors", Jap. J. Appl. Phys., 6, 274 (1967).
1406. Nakahara, K., Tarui, Y., Kawashiro, S., Narukami, N. and
      Hayashi, Y., "Methods of measuring surface potentials of an
      IC by using the electron beam", Bull. Electrotech. Lab., 26
      (4-5), 140-6 (1976), Japanese, EEA79-46344.
1407. Olshanskii, N. A., Vinokurov, V. A. and De, S. K., "Trans-
      verse shrinkage in electron-beam welding", Weld. Prod., 23
      (12), 1-3 (1976).
1408. Panzer, S. ahd Henneberger, J., "Electron beam machining in
      hybrid technology", Feingeraetetechnik, 27 (5), 209-10 (1978),
      German, EEA81-45764.
1409. Sommerkamp, P. and Heinz, B., "Alloy coating with the jumping
      electron beam", Finommech. & Mikrotech., 15 (8), 232-5 (1976),
      Hungarian, EEA80-3295.
1410. Spector, C. J., "Real-time sensor for localized pressure in
      e-beam machining", IBM Tech. Disclosure Bull., 19 (8), 3121
      (1977), EEA80-45643.
1411. Varnell, G. L., Williamson, R. A., Brewer, T. L., Bartelt, J.
      L. and Brown, G. A., "IC fabrication using electron-beam
      technology", Report AD-A056717/2SL, Texas Instruments, Inc.,
      Dallas, 1978, 57 pp.
1412. Zubeck, R. B., King, C. N., Moore, D. F., Barbee, T. W., Jr.,
      Hallak, A. B., Salem, J. and Hammond, R. H., "Growth morph-
      ologies of thick films of $Nb_3Sn$ formed by electron beam evap-
      oration", Thin Solid Films, 40 (1-2-3), 249-61 (1977).

XVI.   LASER TECHNIQUES
       1.  Resistor Trimming

1413. Aggarwal, B. K., "Laser personalizable resistors", IBM Tech.
      Disclosure Bull., 21 (3), 1124-5 (1978).
1414. Aggleton, R. G., "Improved laser trimming for close tolerance
      resistors", Circuit World, 3 (3), 18-19 (1977), EEA81-10040.
1415. Brugman, J. M., "Laser functional trimming techniques for
      consumer thick film circuits", International Conference on
      Manufacturing and Packaging Techniques for Hybrid Circuits,
      1976, p. 87-97, EEA79-46399.
1416. Bube, K. R., "Laser-induced microcrack in thick-film resis-
      tors - a problem and solutions", Am. Ceram. Soc. Bull., 54
      (5), 528-31 (1975), EEA79-1026.
1417. Bube, K. R., Miller, A. Z., Howe, A. and Antoni, B., "Influence
      of laser-trim configuration on stability of small thick-film
      resistors", Solid State Technol., 21 (11), 55-60 (1978).
1418. Cadenhead, R. L., "Production trade-offs in laser trimming
      operations", Solid State Technol., 19 (1), 39-42, 46 (1976),
      EEA79-16868.
1419. Cote, R. E., Headley, R. C., Herman, J. T. and Howe, A.,
      "Effect of laser trimming on thick film resistor stability",
      Circuits Manuf., 17 (7), 14-23 (1977), EEA81-19154.
1420. Dow, R., Mauck, M., Richardson, T. and Swenson, E., "Reducing
      post-trim drift of thin-film resistors by optimizing YAG
      laser output characteristics", IEEE Trans. Components, Hybrids
      & Manuf. Technol., CHMT-1 (4), 392-7 (1978).
1421. Gittoes, J. A., "Laser trimming", Electron. Equip. News,
      10-11 (1976), EEA79-12999.
1422. Gittoes, J., "Functional adjustment using laser trimming",
      Electron. Eng., 49 (594), 53-4 (1977), EEA80-35897.
1423. Gregor, E. and Guio, P., Jr., "The pulsed xenon ion laser -
      its place in present day resistor trimming technology", Pro-
      ceedings of the Society of the Photo-Optical Instrumentation
      Engineers, Vol. 86, High Power Laser Technology, 1976,
      p. 105-11, EEA80-24615.
1424. Horstmann, H. W., "Laser trimming of film resistors", Radio
      Mentor Electron., 43 (10), 404 (1977), German, EEA81-9957.
1425. Illyefalvi-Vietz, Z., "Comparison of art erosive and laser
      beam trimming of thin film resistors", Electrocompon. Sci. &
      Technol., 4 (3-4), 179-83 (1977), EEA81-14318.
1426. Isert, H. and Wiltzek, P., "Laser beams in the thick-layer
      technology", Nachr. Telefonbau & Normalzeit, no. 77, 36-41
      (1976), German, EEA79-41579.
1427. Kestenbaum, A., "Laser trimming applications: recent develop-
      ments", OSA/IEEE Conference on Laser and Electrooptical Sys-
      tems, Digest of Technical Papers, 1978, p. 90, EEA81-33015.
1428. Kummer, F. and Taitl, I., "Thermal expansion and laser trim
      stability of ruthenium based thick film resistors", NTG-
      Fachber., no. 60, 28-33 (1977), German, EEA81-23635.

1429. Masopust, O. T., Jr. and Saifi, M. A., "Laser-generated resistor capacitor networks", Western Electric Eng., 21 (2), 48-58 (1977), EEA80-35782.

1430. Mohr, C. L., Ringo, J. A., Baker, R. A. and Stevens, E. H., "Fundamental limitations on the laser trimming of active networks", Proc. IEEE, 65 (2), 269-71 (1977), EEA80-9190.

1431. Oakes, M., "An introduction to thick film resistor trimming by laser", Opt. Eng., 17 (3), 217-24 (1978), EEA81-45761.

1432. Price, J. J., "A passive laser-trimming technique to improve D/A linearity", 1976 IEEE International Solid-State Circuits Conference, Digest of Technical Papers, p. 104-5, EEA79-46462.

1433. Richardson, T., "Laser trimmer set-up: how to maintain high throughput from lot-to-lot", Circuits Manuf., 16 (9), 80-1, 83 (1976), EEA80-9198.

1434. Rosen, H. G., "Precision laser trimming of thick film circuits in large-scale quantity production", Laser 75 Opto-Electronics Conference, Proceedings, 1975, p. 103-4, Publ. 1976, EEA80-3301.

1435. Schmidling, J., "Laser trimmer branches out", Electro-Opt. Syst. Des., 7 (7), 54-5 (1975).

1436. Shah, J. S. and Berrin, L., "Mechanism and control of post-trim drift of laser-trimmed thick-film resistors", IEEE Trans. Components, Hybrids & Manuf. Technol., CHMT-1 (2), 130-6 (1978), EEA81-33025.

1437. Shibut, D., Conradt, R. and Antoni, W., "Optimizing laser trimming throughput", Circuits Manuf., 18 (10), 30, 32 (1978).

1438. Ujvari, B., "Laser applications in the production of electronics components", Finommech. & Mikrotech., 14 (10), 308-16 (1975), Hungarian, EEA79-13002.

1439. Ulbrich, W., "Resistor geometry comparison with respect to current noise and trim sensitivity", Electrocomponent Sci. Technol., 4 (2), 63-8 (1977).

1440. Waite, G. C., "Laser application to resistor trimming", Proceedings of the Society of the Photo-Optical Instrumentation Engineers, Vol. 86, High Power Laser Technology, 1976, p. 97-101, EEA80-24608.

## 2. Hole Drilling

1441. Clark, K. and Shapiro, M. R., "Laser piercing of materials by induced shock waves", Patent USA 4115683, Publ. September 1978.

1442. Cocca, T. and Dakesian, S., "Laser drilling of vias in dielectric for high density multilayer thick film circuits", Solid State Technol., 21 (9), 63-6 (1978), EEA82-644.

1443. Cohen, M. G. and Kaplan, R. A., "Laser machining of micron-dimension rectangular holes on thin film photomasks", Conference on Laser and Electrooptical Systems, Digest of Technical Papers, 1976, p. 8, EEA80-9505.

1444. Hamilton, D. C. and James, D. J., "Hole drilling with a repet-
      itively-pulsed TEA CO$_2$ laser", J. Phys. D, 9 (4), L41-L43
      (1976).
1445. Howrilka, F. and Lo, J. C., "Hole cleaning", IBM Tech. Dis-
      closure Bull., 21 (3), 961 (1978).
1446. Takaoka, T., Nagano, Y. and Sumiya, M., "Laser machining
      takes on practical dimensions in non-contact, localized
      material processing", JEE, no. 117, 24-7 (1976), EEA79-44935.
1447. Uglov, A. A., Kokora, A. N. and Orekov, N. V., "Laser drilling
      of holes in materials with different thermal properties",
      Sov. J. Quantum Electron., 6 (3), 311-15 (1976), PA79-78369.
1448. Uribe, F., "Laser contoured and drilled substrates for thick
      film applications", Report BDX-613-1563, Bendix Corp., Kansas
      City, Mo., 1976, 20 pp.
1449. Young, A., "Welding and drilling with pulsed lasers", Met.
      Constr., 10 (1), 34-5 (1978).

### 3. Others

1450. Fletcher, M. J., "Laser cutting", Eng. Mater. and Des., 21
      (6), 52-3 (1977), EEA81-2032.
1451. Iwata, A., Nakayama, S., Saito, Y. and Tansho, K., "Laser
      functional trimming for hybrid integrated RC active filters",
      Trans. Inst. Electron. & Commun. Eng. Jpn. Sect. E, E60 (11),
      667-8 (1977), EEA81-14345.
1452. Jaerisch, W. and Makosch, G., "Interferometric surface mapping
      with variable sensitivity [IC manufacturing application]",
      Appl. Opt., 17 (5), 740-3 (1978), EEA81-28173.
1453. Kirby, P. L., "Laser scribing of ceramic substrates", Ceram.
      Ind., 109 (6), 32-4 (1977).
1454. Klauser, H. E., "Dual-function laser system", IBM Tech. Dis-
      closure Bull., 20 (3), 959 (1977), EEA81-19140.
1455. Kubo, L. Y., "YAG laser cutting of c-axis sapphire", Proceed-
      ings of the 28th Electronic Components Conference, 1978,
      p. 85-6, EEA81-37904.
1456. Levinson, G. R. and Smilga, V. I., "Laser processing of thin
      films", Sov. J. Quantum Electron., 6 (8), 885-97 (1976),
      EEA80-14109.
1457. North, J. C., "Laser vaporization of metal films - effect of
      optical interference in underlying dielectric layers", 1976
      International Electron Devices Meeting, Technical Digest,
      p. 614-17, EEA80-41426.
1458. Paek, U. C. and Zeleckas, V. J., "Scribing the mini-substrates
      using CO$_2$ and YAG lasers", Plat. & Surf. Finish., 63 (1),
      60-2 (1976), EEA79-20511.
1459. Platakis, N. S., "Mechanism of laser-induced metal-semiconduc-
      tor electrical connections in MOS structures", J. Appl. Phys.,
      47 (5), 2120-8 (1976), EEA79-33124.

1460. Schiffer, F. and Ziermann, R., "Technological investigations concerning the cutting of glass by means of carbon—dioxide laser", Feingeraetetechnik, 27 (2), 61-3 (1978), German, EEA81-42856.

1461. Schonebeck, G. and Waidelich, W., "A laser cutting tool for measuring out geometrical shapes", Laser 77 Opto-electronics, 1977, p. 615-21, German, EEA81-22702.

1462. Serchuk, A., "Lasers open new packaging horizons", Mod. Packag., 51 (1), 29-30 (1978).

1463. Smith, J. G. and Oldham, H. E., "Laser testing of integrated circuits", IEEE J. Solid-State Circuits, SC-12 (3), 247-52 (1977), EEA80-24009.

1464. Straseheg, F. F., "Laser material processing in the electronics industry", Elektron. Anz., 8 (6), 149-53 (1976), German, EEA79-36695.

1465. Takaba, T., Eguchi, N. and Takahara, M., "A new ceramic board exposure system using laser-beam and linear motors", NEC Res. & Dev., no. 41, 8-18 (1976), EEA79-41604.

1466. Tomlinson, J., "The production of microwave integrated circuits by laser machining", Radio & Electron. Eng., 48 (1-2), 43-6 (1978), EEA81-19166.

XVII.  PACKAGE DESIGN
     1.  Computer-Aided Design (CAD)

1467. Anon., "Advanced computer program aids integrated circuit design", Des. Eng., 13 (1978), EEA81-42130.

1468. Beke, H., Mazur, S., Govaerts, R., Sansen, W. and Van Overstraeten, R., "CALHYM: a computer program for the automatic layout of large digital hybrid microcircuits", Electrocompon. Sci. & Technol., 4 (3-4), 151-6 (1977), EEA81-19164.

1469. Bencsath, p. and Szekely, V., "The survey of the 'Tranztran' circuit analysis program and the application in hybrid circuits design", Hiradstech. Ipari Kut. Intez. Koezl., 17 (1), 41-50 (1977), Hungarian, EEA81-23184.

1470. Braun, J. and Hala, R., "Searching for connections on circuit plates with the aid of a computer", Slaboproudy Obz., 37 (4), 181-6 (1976), Czech, EEA79-28614.

1471. Brinkmann, K. D. and Mynski, D. A., "Computer-aided chip minimization for IC-layout", Proceedings of the 1976 IEEE International Symposium on Circuits and Systems, p. 650-3, EEA80-3529.

1472. Brinsfield, J. G. and Tarrant, S. R., "Computer aids for multilayer printed wiring board design", 12th Design Automation Conference, Proceedings, 1975, p. 296-305.

1473. Burkhardt, W. and Eschke, W., "Computer-aided design of connection for printed circuits in mainly discrete circuit elements", Fernmeldetechnik, 18 (3), 86-7 (1978), German, EEA81-49485.

1474. Cherian, G., "Solving the design problems of a Zif connector to accept leaded LSI packages", Proceedings of the 13th Electrical/Electronics Insulation Conference, 1977, p. 117-23, EEA81-14241.

1475. Christley, F. M., "Thick film substrate (micropackage) design utilizing interactive computer aided design systems", 14th Proceedings of Design Automation Conference, 1977, p. 450-9.

1476. Fairbairn, D. G. and Rowson, J. A., "ICARUS: an interactive integrated circuit layout program", Proceedings of the Fifteenth Annual Design Automation Conference, 1978, p. 188-92, EEA81-42133.

1477. Fukahori, K. and Gray, P. P., "Computer simulation of integrated circuits in the presence of electrothermal interaction", IEEE J. Solid-State Circuits, SC-11 (6), 834-46 (1976), EEA80-6678.

1478. Gavrilov, M. A. and Allan, J., "Design automation systems (with special reference to discrete microelectronic units)", Proceedings of the IFIP Working Conference on Computer-Aided Design Systems, 1976, p. 267-88.

1479. Gavrilov, M. A. and Gilchrist, B., "Automation of electronic and microelectronic design", Information Processing 77, 1977, p. 757-75.

1480. Grigorev, V. V., Kaplan, L. M., Krol, S. M. and Kushnir, B. I., "On the computer-aided design of component topology and automated production of photoprints for microwave hybrid ICs", Poluprovodn. Tekh. & Mikroelektron., no. 24, 86-9 (1976), Russian, EEA80-509.

1481. Gwyn, C. W., "Integrated set of computer aids for custom IC design", Report SAND-76-9216, Sandia Labs., Albuquerque, N. Mex., 1976, 24 pp.

1482. Hadlock, F., "PRIMS. An interactive system for the design of hybrid circuit layout", Comput. Graphics, 10 (2), 293-301 (1976).

1483. Hewlett, R., "Software partners hardware for efficient and faster CAD", Electron, no. 126, 27-8 (1977), EEA81-23578.

1484. Hickman, H. H., Kalke, A. R., Lach, W. and Norman, L. J., "Hybrid integrated circuit assembly using computerized image analysis", Proceedings of the 28th Electronic Components Conference, 1978, p. 356-61, EEA81-42163.

1485. Hosseini, N. M. and Shurmer, H. V., "Computer-aided design of microwave integrated circuits", Radio & Electron. Eng., 48 (1-2), 85-8 (1978), EEA81-19069.

1486. Ikemoto, Y., Sugiyama, T., Igarashi, K. and Kano, H., "Correction and wiring check-system for master-slice LSI", Proceedings of the 13th Annual Design Automation Conference, 1976, p. 336-43, EEA79-37636.

1487. Infante, B., Bracken, D., McCalla, B., Yamakoshi, S. and Cohen, E., "An interactive graphics system for the design of integrated circuits", Proceedings of the Fifteenth Annual Design Automation Conference, 1978, p. 182-7, EEA81-42132.

1488. Jamrich, V., "Several possibilities of using programmable cal-
      culators for the design of electronic circuits", Sdelovaci
      Tech., 25 (1), 21-4 (1977), Czech, EEA80-20248.
1489. Kani, K., Kawanishi, H. and Kishimoto, A., "ROBIN: a building
      block LSI routing program", Proceedings of the 1976 IEEE
      International Symposium on Circuits and Systems, p. 658-61,
      EEA80-3552.
1490. Koller, K., Kubosch, W. and Lauther, U., "Computer-aided
      design of MOS-LSI circuits using the AVESTA program system",
      Siemens Forsch.- & Entwicklungsber., 5 (6), 350-2 (1976),
      EEA80-3547.
1491. Kreinberg, W., "Computer-aided development of large integrated
      circuits using interactive systems", Elektron. Anz., 8 (5),
      122-3 (1976), German, EEA79-33110.
1492. Krivosheikin, A. V., "Calculation of component tolerances of
      microcircuits using cost as criterion", Izv. VUZ Radioelek-
      tron., 19 (6), 108-12 (1976), Russian, EEA79-46349.
1493. Larson, R., "Computer-engineer partnerships produce precise
      layouts fast", Electronics, 51 (2), 102-7 (1978), EEA81-14247.
1494. Mahapatra, S. and Bakir, Q. H., "Computer aided design and
      evaluation of MIC hybrid coupled pin diode phase shifters",
      IEE-IERE Proc. India, 16 (1), 19-28 (1978), EEA81-37482.
1495. Martin, R. E., Shah, R. R. and Smith, M. T., "Computer-aided
      design of hybrid microcircuits: Four years of operating ex-
      perience", Proceedings of the 1977 International Microelec-
      tronics Symposium, p. 36-9.
1496. Maruhashi, T. and Nakaoka, M., "Analog-hybrid simulation of
      thyristor circuit and its application for optimal design",
      Mem. Fac. Eng. Kobe Univ., no. 23, 115-22 (1977), EEA81-13654.
1497. Meindl, J. D., Saraswat, K. C., Dutton, R. W. and Gibbons,
      J. F., "Computer aided engineering of semiconductor inte-
      grated circuits", Report AD-A056914/5SL, Stanford Univ.,
      Calif., Integrated Circuits Lab., 1978, 89 pp.
1498. Myers, R. L., "Computer aided design of printed circuit
      boards", Circuit World, 4 (2), 10-11 (1978), EEA81-37876.
1499. Nishioka, I., Suzuki, C. and Okano, K., "Minicomputerized
      layout system speeds up circuit patterns for PC boards",
      JEE, no. 134, 50-4 (1978), EEA81-37877.
1500. Odawara, G., Iio, J. and Kazuhiko, K., "Study on the circuit
      and packaging design aided by computer", J. Fac. Eng. Univ.
      Tokyo Ser. B, 33 (4), 481-502 (1976), EEA80-24250.
1501. Otten, R. H. J. M. and van Lier, M. C., "Automatic IC-layout:
      the geometry of the islands", 1975 IEEE International Sympos-
      ium on Circuits and Systems, p. 231-4, EEA79-45355.
1502. Persky, G., Deutsch, D. N. and Schweikert, D. G., "LTX - a
      minicomputer-based system for automated LSI layout", J. Des.
      Autom. & Fault-Tolerant Comput., 1 (3), 217-55 (1977),
      EEA80-32460.

1503. Preas, B. T. and Gwyn, C. W., "Methods for hierarchical auto-
      matic layout of custom LSI circuit masks", Proceedings of
      the Fifteenth Annual Design Automation Conference, 1978,
      p. 206-12, EEA81-46049.
1504. Price, I. R. and Moran, P., "Wiring algorithms for the com-
      puter-aided layout of multilayer hybrid microcircuits", IEEE
      Trans. Parts, Hybrids & Packag., PHP-13 (2), 143-6 (1977),
      EEA80-32249.
1505. Puri, P., "A unique approach for generating tester programs
      for LSI design", Digest of Papers 1978 Semiconductor Test
      Conference, p. 218-26.
1506. Ramotowski, M., "Automatically aided design of electronic
      circuits by improved NAP-2 programme", Elektronika, 19 (4),
      174-7 (1978), Polish, EEA81-41811.
1507. Ripka, G. and Albrecht, M., "Small computer aided design and
      documentation of thick film integrated circuits. I.", Finom-
      mech. & Mikrotech., 14 (9), 269-74 (1975), Hungarian,
      EEA79-8984.
1508. Ripka, G. and Albrecht, M., "Small computer aided design and
      documentation of thick film integrated circuits. II.", Finom-
      mech. & Mikrotech., 14 (12), 375-80 (1975), EEA79-16866.
1509. Rychkov, L. A. and Alekseeva, R. P., "Automatic routing of
      connections on multilayer printed circuit boards", Prib. &
      Sist. Upr., no. 9, 7-9 (1977), Russian, EEA81-14254.
1510. Ryzko, J. and Syta, A., "Application of electronic digital
      computers to DC analysis of elementary cells of MOS integrated
      circuits", Arch. Elektrotech., 25 (2), 477-9 (1976), Polish,
      EEA80-21016.
1511. Shah, P., "Computer aided design and optimization for semi-
      conductor device fabrication", Electrochemical Society
      Spring Meeting, Extended Abstracts, 1977, p. 589-91,
      EEA81-19405.
1512. Sugiyama, Y., "Taneda, M., Ueda, K., Kani, K. and Teramoto,
      M., "COMPAS: multichip LSI routing program", Electr. Commun.
      Lab. Tech. J., 25 (6), 969-83 (1976), Japanese, EEA79-45323.
1513. van Lier, M. C. and Otten, R. H. J. M., "Automatic IC-layout:
      some planarization algorithms", Proceedings of the 1976 IEEE
      International Symposium on Circuits and Systems, p. 662-5,
      EEA80-3530.
1514. Vasilchenko, V. I., Gitman, I. R., Zgonnik, O. N., Krivo-
      sheikin, A. V., Miksturov, N. N., Stupachenko, A. A. and
      Flekser, L. A., "Computer-aided design of composite integrated
      microcircuits", Izv. VUZ Radioelektron., 18 (6), 9-17 (1975),
      Russian, EEA79-1001.
1515. Villers, P., "A minicomputer based interactive graphics system
      as used for electronic design and automation", Proceedings
      of the Fifteenth Annual Design Automation Conference, 1978,
      p. 446-53, EEA81-45717.

1516. Waldvogel, C. W., "How to build a multilayer hybrid substrate using computer generated artwork and thick film technology", Insul./Circuits, 22 (12), 31-4 (1976), EEA80-9203.

1517. Yamamoto, K., "Design of irredundant MOS network - a program manual for the design algorithm DIMN (Design of Irredundant MOS Network)", Report R-76-784, Illinois Univ., Urbana, 1976, 125 pp.

1518. Zaitseva, Z. N. and Shtein, M. E., "Hamiltonian cycles in layout problems", Cybernetics, 12 (3), 405-8 (1976).

1519. Zibert, K., "A computer-aided method for laying out hybrid circuits", NTG-Fachber., 51, 125-32 (1975), German, EEA79-28708.

1520. Zvenigorodskii, E. G. and Matveev, V. I., "Automated design of hybrid film microcircuits", Prib. and Sist. Upr., no. 11, 12-13 (1977), Russian, EEA81-18726.

## 2.  Others

1521. Ackermann, W., "Components of hybridizing film circuits", Components Rep., 10 (3), 95-8 (1975), EEA79-9026.

1522. Ackermann, W. and Reinl, H., "Do you design your film circuit for fabrication?", Components Rep., 10 (5), 134-8 (1975), EEA79-33076.

1523. Advani, L. R., "Analyzing complex packaging systems through simulation", Package Dev., 6 (4), 13-17 (1976).

1524. Ahmed, K. U., "Solvable nonuniform distributed parameter Y-Z-KY networks", IEE-IERE Proc. India, 15 (3), 111-16 (1977), EEA80-41108.

1525. Alexandersson, R. and Rorstrom, H. O., "New packaging structure for electronic switching equipment", Ericsson Rev., 53 (2), 100-7 (1976).

1526. Alia, G., Ciompi, P., Martinelli, E. and Bernardini, F., "LSI components modelling in a three-valued functional simulation", Proceedings of the Fifteenth Annual Design Automation Conference, 1978, p. 428-38, EEA81-46050.

1527. Asahara, S., Tsukiyama, S., Shirakawa, I. and Ozaki, H., "An approach for the via assignment problem in backboard wiring", Trans. Inst. Electron. & Commun. Eng. Jpn. Sect. E, E61 (3), 244-5 (1978), EEA81-37878.

1528. Asano, T., Kitahashi, T. and Tanaka, K., "On a method of realizing minimum-width wirings", Electron. Commun. Jpn., 59 (2), 29-39 (1976), EEA80-41522.

1529. Asano, T., Kitahashi, T., Tanaka, K., Horino, H. and Amano, N., "A wire-routing scheme based on trunk-division methods", IEEE Trans. Comput., C-26 (8), 764-72 (1977), EEA81-904.

1530. Balde, J. W. and Amey, D. I., "New chip carrier package concept", Computer, 10 (12), 58-68 (1977).

1531. Bechtolsheim, A., "Interactive specification of structured design", Proceedings of the Fifteenth Annual Design Automation Conference, 1978, p. 261-3, EEA81-45734.

1532. Berndlmaier, E. and Dorler, J. A., "High performance package", IBM Tech. Disclosure Bull., 20 (8), 3090-1 (1978).

1533. Beyer, M., "Determination of the efficacy in the choice of methods for making junctions and connections", Elektrie, 30 (10), 518-20 (1976), German, EEA80-5816.

1534. Breuer, M. A., "Min-cut placement", J. Des. Autom. & Fault-Tolerant Comput., 1 (4), 343-62 (1977), EEA81-5601.

1535. Brinkmann, K. D., "The chip net - a powerful aid to design MSI- and LSI-layout", Nachrichtentech. Z. (NTZ), 30 (12), 929-35 (1977), German, EEA81-19519.

1536. Brinkmann, K. D. and Mlynski, D. A., "Layout-partitioning method for MSI and LSI", 1977 IEEE International Symposium on Circuits and Systems Proceedings, p. 160-14, EEA80-32467.

1537. Brooks, P., "Designer's guide to physical design and packaging - 1, 2, 3", EDN, 21 (18, 19, 20), 46-52, 99-103, 75-80 (1976).

1538. Cardwell, W. T., Jr., "IC bricklaying for miniature projects", Radio-Electron., 48 (12), 58-60 (1977), EEA81-28172.

1539. Carey, B. J. and Grossman, H., "Assembling a complex breadboard can be as easy as 1, 2, 3", Electronics, 50 (20), 104-9 (1977).

1540. Cherian, G., "Solving the design problems of a Zif connector to accept leaded LSI packages", Proc. of the Electr./Electron. Insul. Conf., 13th, 1977, p. 117-23.

1541. Chow, W. C., "Packaging tradeoffs for high performance computer", 18th IEEE Computer Society International Conference, Digest of Papers, 1979, p. 290-4.

1542. Coles, J., "The role of the hybrid microcircuit in professional electronics", New Electron., 10 (6), 66, 69, 71 (1977), EEA80-35933.

1543. Crampagne, R. and Khoo, G., "Synthesis of certain transmission lines employed in microwave integrated circuits", IEEE Trans. Microwave Theory & Tech., MIT-25 (5), 440-2 (1977), EEA80-17043.

1544. Crowther, G. O. and Emms, E. T., "Inter-relation between LSI technology and electronic system design", Tijdschr. Ned. Elektron.- and Radiogenoot., 41 (5), 165-9 (1976), Dutch, EEA80-21006.

1545. Danzig, A. C., "Methods for achieving high density circuit patterns", Insul./Circuits, 21 (9), 36-7 (1975).

1546. Deboskey, W. R. and Rajac, T. J., "Encapsulated lead frame - A package design for mechanical thermal pulse bonding", Insul./Circuits, 22 (2), 35-7 (1976), EEA79-24054.

1547. Deutsch, D. N., "A 'Dogleg' channel router", Proceedings of the 13th Annual Design Automation Conference, 1976, p. 425-33, EEA79-37639.

1548. Director, S. W. and Hachtel, G. D., "Simplicial approximation for design centering", IBM Tech. Disclosure Bull., 19 (4), 1490-3 (1976), EEA80-9187.

1549. Doreau, M. T. and Abel, L. C., "A topologically based non-minimum distance routing algorithm", Proceedings of the Fifteenth Annual Design Automation Conference, 1978, p. 92-9, EEA81-42123.

1550. Ducamus, J., "Topology and implantation in monolithic integrated circuit design", Onde Electr., 57 (4), 300-4 (1977), French.

1551. Dutton, R. W., Gonzalez, A. G., Rung, R. D. and Antoniadis, D. A., "IC process engineering models and applications", Electrochemical Society Spring Meeting, Extended Abstracts, 1977, p. 585-8, EEA81-19404.

1552. Dydyk, M., "Master the T-junction and sharpen your MIC designs", Microwave, 16 (5), 184-6 (1977), EEA81-10051.

1553. Eberlein, D. D., "Custom MSI for very high speed computers", 18th IEEE Computer Society International Conference, Digest of Papers, 1979, p. 295-8.

1554. Etiemble, D., "TTL circuits for a 4-valued bus. A way to reduce package and interconnections", Proceedings of the Eighth International Symposium on Multiple-Valued Logic, 1978, p. 7-13, EEA81-37648.

1555. Forster, J. H., "Polychip DIPs: quick-turnaround custom circuit design [telecommunication equipment]", Bell Lab. Rec., 55 (8), 208-14 (1977), EEA81-131.

1556. Fox, B. C. and Kraft, W. R., "Incremental masterslice part number design", IBM Tech. Disclosure Bull., 20 (3), 1116-19 (1977), EEA81-24035.

1557. Franson, P., "Don't overlook semicustom ICs for your next design project", EDN, 22 (3), 72-6 (1977), EEA80-41523.

1558. Frasso, V. J., "Applying structural/dynamic analysis to package design", Electron. Packag. & Prod., 18 (3), 84-6 (1978).

1559. Gianfagna, M. A., "A unified approach to test data analysis", Proceedings of the Fifteenth Annual Design Automation Conference, 1978, p. 117-24, EEA81-42131.

1560. Gibson, D. and Nance, S., "SLIC - symbolic layout of integrated circuits", Proceedings of the 13th Annual Design Automation Conference, 1976, p. 434-40, EEA79-37640.

1561. Gniewek, J. J. and Kelly, J. H., "Via design to minimize stress levels in multilayer ceramic substrates", IBM Tech. Disclosure Bull., 20 (11), 4774-5 (1978).

1562. Gold, R. D., "Material selection in hybrid production design", Solid State Technol., 20 (1), 31-5, 37 (1977).

1563. Goto, S. and Kuh, E. S., "An approach to the two-dimensional placement problem in circuit layout", IEEE Trans. Circuits & Syst., CAS-25 (4), 208-14 (1978), EEA81-32993.

1564. Hart, W. S. and Jones, N. G., "Nonsymmetrical design for resistor module pack", IBM Tech. Disclosure Bull., 21 (3), 966 (1978).

1565. Heller, W. R., Mikhail, W. F. and Donath, W. E., "Prediction of wiring space requirements for LSI", Proceedings of the 14th Annual Design Automation Conference, 1977, p. 32-42.

1566. Hinow, C., Hinow, G., Mitew, M. and Georgiew, D., "On a methodology for wiring electronic devices", Fernmeldetechnik, 18 (3), 92-3 (1978), German, EEA81-49489.

1567. Ho, H. K., "Automatic wire-bonding and wire-removal", IBM Tech. Disclosure Bull., 19 (8), 3043-5 (1977), EEA80-40320.

1568. Hou, H. S., "Application of uniform loading theory to circuit packaging and memory arrays in high-speed computers", IEEE Trans. Comput., C-21 (5), 454-63 (1972), CCA7-12409.

1569. Hung, R. Y. and Levine, J. L., "Modular approach on package design for roled display", IBM Tech. Disclosure Bull., 19 (12), 4805 (1977), EEA81-16666.

1570. Hutchins, W. D., "Microwave design utilization with a new nonceramic high dielectric substrate", Proceedings of the 1977 International Microelectronics Symposium, p. 175-80.

1571. Ilin, V. N., "Techniques for optimising the accuracy of static mathematical models of electronic circuit components", Izv. Vuz Radioelektron., 20 (3), 60-4 (1977), Russian, EEA80-28847.

1572. Irlin, A. V. and Kempa, Y. M., "Problems of integrated micro-miniaturisation of HF devices", Poluprovodn. Tekh. & Mikroelektron., no. 24, 4-20 (1976), Russian, EEA80-507.

1573. Ishii, T., Kondoh, A. and Shirahata, K., "A new package for negative resistance microwave diodes", 5th European Microwave Conference, 1975, p. 710-14, EEA79-4097.

1574. Johnson, W. E., Mathad, G. S. and Millman, D. S., "Control algorithm for forming zero overlap via holes (LSI applications)", IBM Tech. Disclosure Bull., 19 (3), 878-82 (1976), EEA80-6680.

1575. Joyce, B. T., "A circuit designer meets hybrid technology", IEEE Trans. Manuf. Technol., MFT-6 (4), 73-80 (1977), EEA81-14342.

1576. Jumbeck, G. I. and Kriger, R. J., "LSI chip design for ease of engineering change", IBM Tech. Disclosure Bull., 20 (7), 2632 (1977), EEA81-46037.

1577. Kanda, Y. and Migitaka, M., "Design considerations for Hall devices in Si IC", Phys. Status Solidi A, 38 (1), K41-4 (1976), EEA80-6687.

1578. Kato, H., Kawanishi, H., Goto, S., Oyamada, T. and Kani, K., "On automated wire routing for building-block MOS LSI", Proceedings of the 1974 IEEE International Symposium on Circuits and Systems, p. 309-13, EEA77-38893.

1579. Keiser, G. E. and Lesinski, L. C., "Multiple-pin connector with internal surge protection", IEEE Trans. Electromagn. Compat., EMC-20 (1), 211-15 (1978).

1580. Khadpe, S., "Design and cost considerations for ceramic chip carrier packaging of active devices", Proceedings of the 1977 International Microelectronics Symposium, p. 270-3.

1581. Kim, M. J., "Design problem of linear IC's", J. Korean Inst. Electron. Eng., 13 (3), 95-100 (1976), Korean, EEA80-682.

1582. Kirby, P., "Thick film advances simplify complex hybrid module design", Electron. Engineering, 48 (577), 35-8 (1976), EEA79-16905.

1583. Kocsis, A., "A thick film hybrid pocket multimeter", International Conference on Manufacturing and Packaging Techniques for Hybrid Circuits, 1976, p. 209-17, EEA79-46572.

1584. Kozel, J. T., "Programmable array module", IBM Tech. Disclosure Bull., 21 (3), 1103-4 (1978).

1585. Krause, G., "Extremely high packaging densities by operating integrated circuits with RF power", Electron. Lett., 11 (15), 327-8 (1975).

1586. Krivosheykin, A. V., "Optimal synthesis of the tolerances on the elements of thin-film hybrid circuits", Radio Eng. & Electron. Phys., 21 (11), 93-9 (1976), EEA81-913.

1587. Leffler, D., "Parameters influencing the optimisation of the leading dimensions of single-layer circuit boards", Fernmeld-etechnik, 18 (3), 84-5 (1978), German, EEA81-49484.

1588. Lew, J. S., "Method for designing efficient connector patterns for mounting chips in LEM technology", IBM Tech. Disclosure Bull., 18 (7), 2390-3 (1975), EEA79-12987.

1589. Libove, C., "Rectangular flat-pack lids under external pressure: formulas for screening and design", 13th Annual Proceedings of Reliability Physics Symposium, 1975, p. 38-47, EEA79-44930.

1590. Lindner, H., "Design of printed circuit boards with respect to mechanical strength under dynamic loading", Fernmelde-technik, 18 (3), 87-8 (1978), German, EEA81-49486.

1591. Lyman, J., "Growing pin count is forcing LSI package changes", Electronics, 50 (6), 81-91 (1977).

1592. Magrupov, T. M., "A method of routing the connections of single-chip large-scale integrated circuits", Izv. Akad. Nauk UzSSR Ser. Tekh. Nauk, no. 6, 13-16 (1976), Russian, EEA80-21010.

1593. Malek, T. and Schwartz, W. B., "IC packaging panels for high-speed logic applications", Electron. Packag. & Prod., 16 (4), 35-40 (1976).

1594. Matthews, P., "Towards a custom IC", Electron, no. 121, 43-4 (1977), EEA81-19510.

1595. Meehan, J. W., "Meeting today's detector challenge", Opt. Spectra, 10 (8), 49-51 (1976), EEA80-6951.

1596. Mochalov, B. V. and Ermolaev, Y. P., "Arrangement of units in electronic equipment", Izv. VUZ Radioelektron., 18 (11), 10-14 (1975), Russian, EEA79-11900.

1597. Mueller, V. W. and Valente, A. M., "No. 1 EAX network component redesign and packaging reconfiguration", GTE Autom. Electr. Tech. J., 14 (7), 314-19 (1975).

1598. Nakahara, H., Burr, R. P. and Burr, J. B., "Interconnection routing for multiwire circuit boards", Proceedings of the 1976 IEEE International Symposium on Circuits and Systems, 1976, p. 646-9, EEA80-3281.

1599. Neumann, K. and Becker, R., "Modular systems for constructing electronic equipment", Feingeraete Tech., 24 (10), 467-8 (1975), EEA79-7835.

1600. Newman, P. C., "More and more on less and less", Electron. & Power, 22 (7), 444-8 (1976), EEA79-37615.

1601. Nishioka, I., Kurimoto, T., Yamamoto, S., Shirakawa, I. and Ozaki, H., "An approach to gate assignment and module placement for printed wiring boards", Proceedings of the Fifteenth Annual Design Automation Conference, 1978, p. 60-9, EEA81-42119.

1602. Olivieri, D., "Don't overlook the package design", Microwaves, 14 (11), 32-4, 36-7, 72 (1975), EEA79-15888.

1603. Osborne, R. J., "Can mechanical engineers design electronic circuits? I.", Chart. Mech. Eng., 23 (9), 71-2, 74-6 (1976), EEA80-13276.

1604. Parisi, J., "Decoupling capacitor placement", IBM Tech. Disclosure Bull., 19 (8), 3046-7 (1977), EEA80-41117.

1605. Peck, P. E., "Functional design benefits maintenance", Executive Eng., 56 (9-10), 7-10 (1976), EEA80-8379.

1606. Peirce, C., "Hybrid circuits provide design flexibility", Electron, no. 84, 15, 27 (1975), EEA79-9033.

1607. Petrikovits, L. and Goblos, I., "Reliability testing of thick film circuit models", Hiradastechnika, 28 (4), 101-9 (1977), Hungarian, EEA81-5623.

1608. Pickvance, E., "Custom design perspectives (ICs)", Microelectron. & Reliab., 14 (4), 383 (1975), EEA79-12993.

1609. Pollack, H. W., "For simple, compact packaging, use flex circuitry and flat cable", Insul./Circuits, 23 (9), 53-5 (1977), EEA80-40323.

1610. Popov, V. P., Ivanstov, V. P. and Chernov, I. V., "Design-cost characteristics of large-scale hybrid integrated circuits", Izv. VUZ Radioelektron., 18 (12), 3-10 (1975), Russian, EEA79-13050.

1611. Rau, B. R., "New philosophy for wire routing", Report SU-SEL-75-049, Stanford Univ., Calif., 1975, 42 pp.

1612. Riel, M. and Chalman, R., "Specifying special IC packages", Electron. Packag. & Prod., 15 (5), 65-6, 68 (1975).

1613. Rubin, F., "Comparison of wire spreading metrics for printed circuit routing", J. Des. Autom. Fault-Tolerant Comput., 2 (3), 231-9 (1978).

1614. Ruwe, V. W., Herron, J. A., Sulouff, R. E. and Hartin, S. S., "Engineering considerations in large-scale hybrids", Proceedings of the 1977 International Microelectronics Symposium, p. 32-6.

1615. Sandeau, R. F., "Total integration of microelectronics and electronic package design", Proceedings of the 1977 International Microelectronics Symposium, p. 29-31.

1616. Schalow, R. D., Golart, T. C. and Young, M. A., "On circuit design for timing margin reliability", Proc. Annu. Reliab. Maintainability Symp., 1978, p. 173-78.

1617. Scheel, W. and Walter, H. P., "Investigation into the stages of making wire-wrap connections", Fernmeldetechnik, 17 (4), 132-4 (1977), German, EEA81-9248.

1618. Shirakawa, I., "The state-of-the-art of placement and routing techniques in packaging design", J. Inst. Electron. & Commun. Eng. Jpn., 61 (3), 245-55 (1978), Japanese, EEA81-48751.

1619. Smith, J. A., "New routing technique to reduce interconnection crossings", J. Des. Autom. Fault-Tolerant Comput., 1 (3), 257-86 (1977).

1620. Sommerfeld, E., "A problem-solving algorithm for minimising partitioning connections and for component string in the design of printed circuit boards", Fernmeldetechnik, 18 (3), 91-2 (1978), German, EEA81-49488.

1621. Sosnitskii, A. V. and Shamin, P. A., "Limitation of path positron fixings in the routing of printed circuit connections for microcircuits", Izv. VUZ Radioelektron., 20 (8), 91-5 (1977), Russian, EEA81-10006.

1622. Soukup, J., "Fast maze router", Proceedings of the Fifteenth Annual Design Automation Conference, 1978, p. 100-2, EEA81-42124.

1623. Sredni, J., "Use of power transformations to model the yield of IC's as a function of active circuit area", 1975 International Electron Devices Meeting, Technical Digest, p. 123-5, EEA79-37645.

1624. Styblinski, M. A., "Some results of analytical study of resistor correlations in IC's based on the model of factor analysis", Proc. IEEE Int. Symp. Circuits Syst., 1978, p. 205-6.

1625. Sugiyama, Y. and Kani, K., "A multichip LSI routing method", Electron. & Commun. Jap., 58 (4), 106-14 (1975), EEA80-6431.

1626. Szanto, L., "Placement of elements on the integrated circuit chip for the minimum signal distortion", Elektrotech. Cas., 27 (1), 29-40 (1975), Slovak, EEA79-20524.

1627. Szolnoki, G. and Zaunrith, F., "Packaging system D for data communications and automatic feedback control systems", Siemens Rev., 42 (11), 490-3 (1975), EEA79-4363.

1628. Tarr, M., "Passive components of thick film circuits", New Electron, 10 (6), 90-2 (1977).

1629. Tebbutt, M. A., "System ISEP - an international standard equipment practice for design engineers", Des. Eng., 52-3 (1975), EEA79-3726.

1630. Ting, B. S. and Kuh, E. S., "An approach to the routing of multilayer printed circuit boards", Proceedings of the 1978 IEEE International Symposium on Circuits and Systems, p. 902-11, EEA81-42117.

1631. Tsunashima, E., "Research injects new creativity into high-density packaging", JEE J. Electron. Eng., no. 104, 26-30 (1975).

1632. Uhlmann, R., "Enclosure concept for multichip-hybrid units", Feingeraete Tech., 25 (3), 117-19 (1976), German, EEA79-24924.

1633. Vacca, A. A., "Considerations for high performance LSI applications", 18th IEEE Computer Society International Conference,
Digest of Papers, 1979, p. 278-84.

1634. Vandre, R. H., "Effects of shadows on photocurrent-compensated
integrated circuits", Report TR-0075 (5250-30)-1 SAMSO-TR-75-
114, Aerospace Corp., El Segundo, Calif., 1975, 26 pp.,
EEA79-4777.

1635. Vilkelis, W. V. and Henle, R. A., "Performance vs circuit
packing density", 18th IEEE Computer Society International
Conference, Digest of Papers, 1979, p. 285-9.

1636. Volkl, W., "Mounting system D and its design features", Feinwerktech. & Messtech., 85 (1), 25-9 (1977), German, EEA80-24005.

1637. Webster, R., "Fine line screen printing yields as a function
of physical design parameters", IEEE Trans. Manuf. Technol.,
MFT-4 (1), 14-20 (1975).

1638. Wilkinson, I. G., "Multi-purpose sub-rack system design", New
Electron., 10 (24), 68 (1977), EEA81-23582.

1639. Wilson, E. A., "Packaging a computer by site - a new efficient
approach", ASME Paper No. 77-DE-19, 1977, 5 pp.

1640. Zarwyn, B., "New concept of survivability enhancement with
self-organizing electronic circuits", Report AD-A026168/5SL,
Harry Diamond Labs., Adelphi, Md., 1976, 15 pp.

XVIII.   THERMAL DESIGN
1.   Cooling

1641. Alcorn, G. E. and Lynch, J. R., "Promoting nucleate boiling
of semiconductor devices in a fluorocarbon coolant", IBM
Tech. Disclosure Bull., 20 (4), 1395 (1977), EEA81-23844.

1642. Altoz, F. E., "Thermal design of airborne radars - present
and future", IEEE 1977 Mechanical Engineering in Radar Symposium, p. 110-13.

1643. Anacker, W., "Liquid cooling of integrated circuit chips",
IBM Tech. Disclosure Bull., 20 (9), 3742-3 (1978), EEA82-631.

1644. Anon., "A new approach to electronic equipment cooling",
OEM Des., 1 (15), 47 (1972), CCA8-7430.

1645. Anon., "The cooling of semiconductor components. II.", Elektronik, 26 (12), 73-4 (1977), German, EEA81-13300.

1646. Antonetti, V. W., Chu, R. C., Moran, K. P. and Simons, R. E.,
"Thermal flex connector", IBM Tech. Disclosure Bull., 20 (2),
682 (1977), EEA81-18371.

1647. Antonetti, V. W., Chu, R. C., Moran, K. P. and Simons, R. E.,
"Compliant cold plate cooling scheme", IBM Tech. Disclosure
Bull., 21 (6), 2431 (1978).

1648. Antonetti, V. W., Hwang, U. P., Moran, K. P. and Pascuzzo,
A. L., "Swivel module conduction module", IBM Tech. Disclosure
Bull., 20 (7), 270-1 (1977).

1649. Antonetti, V. W., Liberman, H. E. and Simons, R. E., "Integrated module heat exchanger", IBM Tech. Disclosure Bull.,
20 (11), 4498 (1978).

1650. Antonetti, V. W. and Simons, R. E., "Air jet spot cooling",
      IBM Tech. Disclosure Bull., 19 (12), 4688-9 (1977), EEA81-10036.
1651. Archey, W. B., Audi, R. D., Clark, R. J. and Redmond, R. J.,
      "Liquid encapsulated integrated circuit package", Patent USA
      3999105, Publ. December 1976.
1652. Arnold, A. J., Clark, B. T., Hardin, P. W., Robbins, G. J.
      and Siwy, M. J., "Self contained data processing package",
      IBM Tech. Disclosure Bull., 20 (10), 3895-7 (1978).
1653. Arnold, A. J. and Hwang, U. P., "Thermally enhanced module
      for self-contained data processing package", IBM Tech. Dis-
      closure Bull., 20 (5), 1766-7 (1977), EEA81-32236.
1654. Bailey, N. P., "Direct immersion cooling block", IBM Tech.
      Disclosure Bull., 21 (6), 2437 (1978).
1655. Balderes, D. and Lynch, J. R., "Liquid cooling of a multichip
      module package", IBM Tech. Disclosure Bull., 20 (11), 4336-7
      (1978).
1656. Balderes, D., Lynch, J. R. and Yacavonis, R. A., "Heat dissi-
      pation from IC chips through module package", IBM Tech. Dis-
      closure Bull., 19 (11), 4165-6 (1977), EEA81-10369.
1657. Balderes, D., Moran, K. P. and Nutter, S. W., "Cooling of
      module with stud-mounted cold plate", IBM Tech. Disclosure
      Bull., 20 (6), 2327-8 (1977), EEA82-106.
1658. Bar-Cohen, A., "Immersion and heat-of-fusion thermal control
      techniques for electronic components", Ben Gurion Univ. Negev
      Dep. Mech. Eng. MED Rep., no. 14, 27 pp. (1975).
1659. Baxter, G. K. and Lew, P. Y., "Thermal effect of material
      properties on electronic modules", Proc. Tech. Program Natl.
      Electron. Packag. Prod. Conf., 1977, p. 420-31.
1660. Bellamy, M., "Latest techniques in circuit heat dissipation",
      New Electron., 10 (15), 25-6 (1977), EEA81-9253.
1661. Berndlmaier, E., Clark, B. T., Dorler, J. A., Metreaud, C. G.
      and Robbins, G. J., "Chip packaging structure with enhanced
      cooling", IBM Tech. Disclosure Bull., 20 (5), 1772 (1977),
      EEA81-32238.
1662. Berndlmaier, E. and Dorler, J. A., "Chip suitable for back-
      side cooling", IBM Tech. Disclosure Bull., 19 (11), 4190 (1977).
1663. Berndlmaier, E. and Dorler, J. A., "Semiconductor package with
      improved cooling", IBM Tech. Disclosure Bull., 20 (9),
      3452-3 (1978), EEA82-113.
1664. Bhattacharya, S., Koopman, N. G., Sullivan, E. J. and Weiss,
      L., "Chip cooling with thermal impedance control", IBM Tech.
      Disclosure Bull., 21 (7), 2819-20 (1978).
1665. Bhattacharya, S. and Sullivan, E. J., "Chip cooling package",
      IBM Tech. Disclosure Bull., 21 (2), 613-14 (1978).
1666. Blake, B. E., Druschel, W. O. and Metreaud, C. G., "Packaging
      structure", IBM Tech. Disclosure Bull., 21 (1), 183-4 (1978).
1667. Bravo, H. V. and Bergles, A., "Limits of boiling heat transfer
      in a liquid-filled enclosure", Proceedings of the 1976 Heat
      Transfer and Fluid Mechanics Institute, p. 114-27, EEA80-19691.

1668. Brennan, P. A., Fang, F. F. and Ruehli, A. E., "Integrated circuit with increased air dielectric", IBM Tech. Disclosure Bull., 20 (3), 1237-8 (1977), EEA81-19150.

1669. Brooks, W. T., "Dielectric liquid", Patent German 2608447, Publ. September 1976, CA85-152918.

1670. Brownlow, L. S., "Cable guide", IBM Tech. Disclosure Bull., 20 (11), 4532-3 (1978).

1671. Bukovskaya, O. I. and Kozdova, L. A., "Results of an investigation of the temperature fields of a hybrid integrated circuit", Teplofiz. & Telplotekh., no. 32, 43-7 (1977), Russian, EEA80-41113.

1672. Cadwallader, R. H., Gruendle, D. E., Poulos, L. J., Scalia, L. and Van Slyke, H. F., "Helium leak test fixture [for semiconductor cooling package]", IBM Tech. Disclosure Bull., 20 (7), 2680-1 (1977), EEA81-45177.

1673. Cain, O. J. and Ordonez, J., "Semiconductor module with improved air cooling", IBM Tech. Disclosure Bull., 19 (5), 1802 (1976).

1674. Carminali, E., "Fluid cooling: Greater safety and less space", Tecnol. Elettr., 7-8, 54-61 (1975), Italian, EEA79-89.

1675. Cases, M., Rubin, B. J. and Wu, L. L., "Large scale integration (LSI) cooling technique", IBM Tech. Disclosure Bull., 20 (12), 5169 (1978).

1676. Chu, R. C., "Conduction cooling", IBM Tech. Disclosure Bull., 21 (2), 751 (1978).

1677. Chu, R. C., "Conduction cooling", IBM Tech. Disclosure Bull., 21 (2), 752-3 (1978).

1678. Chu, R. C. and Hwang, U. P., "Dielectric fluidized cooling system", IBM Tech. Disclosrue Bull., 20 (2), 678-9 (1977), EEA81-18369.

1679. Ciancanelli, J. M., "Circuit module with heat transfer", IBM Tech. Disclosure Bull., 19 (7), 2652 (1976), EEA80-32458.

1680. Clark, B. T., "Circuit module with ceramic fin for cooling", IBM Tech. Disclosure Bull., 19 (4), 1336 (1976).

1681. Clark, B. T. and Metreaud, C. G., "Cooling device for multilayer ceramic modules", IBM Tech. Disclosure Bull., 20 (5), 1769-71 (1977), EEA81-37134.

1682. Coughlin, C. P. and Siwy, M. J., "Structural cooling stack concept for module", IBM Tech. Disclosure Bull., 21 (4), 1471-2 (1978).

1683. Cozine, H. and Roberts, L., "Thermoelectrically cooled detectors - another option", SPIE Semin. Proc. Vol. 132, 1978, p. 118-26.

1684. Damm, E. P., Jr., "Convection cooling apparatus", IBM Tech. Disclosure Bull., 20 (7), 2755-6 (1977), EEA81-46039.

1685. Dance, M., "Power semiconductors", Electron. Ind., 4 (4), 41-3, 45, 47-9 (1978), EEA81-28482.

1686. Dessauer, B., Dessauer, R. G. and Hwang, U. P., "Semiconductor Package", IBM Tech. Disclosure Bull., 21 (7), 2822 (1978).

1687. Dombroski, E. L. and Hwang, U. P., "Thermal conduction stud", IBM Tech. Disclosure Bull., 19 (12), 4683-5 (1977).

1688. Dombroski, E. L. and Hwang, U. P., "Multichip cooling plate", IBM Tech. Disclosure Bull., 21 (2), 745-6 (1978).

1689. Doo, V. Y., "High capacity power dissipation semiconductor package", IBM Tech. Disclosure Bull., 21 (5), 1898-9 (1978).

1690. Doo, V. Y. and Johnson, A. H., "Method of effective cooling of a high power silicon chip", IBM Tech. Disclosure Bull., 20 (4), 1436-7 (1977), EEA81-22848.

1691. Doo, V. Y. and Johnson, A. H., "High performance single chip module", IBM Tech. Disclosure Bull., 20 (4), 1438-9 (1977), EEA81-22849.

1692. Doo, V. Y. and Johnson, A. H., "Semiconductor chip cooling package", IBM Tech. Disclosure Bull., 20 (5), 1440-1 (1977), EEA81-22850.

1693. Dorler, J. A. and Ecker, M. E., "Temperature triggerable fluid coupling system for cooling semiconductor dies", IBM Tech. Disclosure Bull., 20 (11), 4386-8 (1978).

1694. Dumaine, G., Lenoine, J. M. and Pottier, B., "Air-cooled module heat dissipation", IBM Tech. Disclosure Bull., 20 (4), 1472 (1977), EEA81-22851.

1695. Durand, R. D., Hwang, U. P., Oktay, S. and Wong, A. C., "Flexible thermal conductor for electronic module", IBM Tech. Disclosure Bull., 20 (11), 4343-4 (1978).

1696. Eckenbach, H., Krumrein, W. and Tandjung, H., "Semiconductor module with heat transfer", IBM Tech. Disclosure Bull., 20 (11), 5203 (1978).

1697. Ehret, P., Haist, F. and Spielman, W., "Multichip packaging", IBM Tech. Disclosure Bull., 14 (10), 3090 (1972).

1698. Flaherty, R. J. and Toups, J. W., "Multilayer electronics package", IBM Tech. Disclosure Bull., 19 (12), 4489-90 (1977), EEA81-9250.

1699. Foit, J., "Determining thermal resistances of cooling elements", Sdelovaci Tech., 24 (4), 135-6 (1976), Czech, EEA79-40774.

1700. Follette, D. T. and Goodman, H. A., "Staggered logic/memory board configuration for improved card cooling", IBM Tech. Disclosure Bull., 20 (7), 2757-8 (1977).

1701. Follette, D. T., Goodman, H. A. and Gupta, O. R., "Controlled frontal bypass configuration for air cooled computer frames", IBM Tech. Disclosure Bull., 21 (6), 2432-3 (1978).

1702. Follette, D. T., Goodman, H. A. and Schulman, M. D., "Module chip site temperature sensing using thermistor imbedded in piston", IBM Tech. Disclosure Bull., 21 (2), 566-8 (1978).

1703. Frieser, R. G., Fugardi, J. F. and Reeber, M. D., "Generating thermal nucleation sites on semiconductor devices", IBM Tech. Disclosure Bull., 19 (10), 3730 (1977).

1704. Frieser, R. G. and Gregor, L. V., "Purification and monitoring techniques for high dielectric liquids", IBM Tech. Disclosure Bull., 19 (8), 3033 (1977).

1705. Gangwar, M. C., "Cooling effect of an intermittent alternating electric field (dielectric liquids)", Indian J. Pure & Appl. Phys., 13 (9), 640-2 (1975), EEA79-9047.

1706. German, G. and Smith, P. W., "Cooling of electronic equipment in relation to component temperature limitations and reliability", AGARD Conference Proceedings, No. 196 on Avionic Cooling and Power Supplies for Advanced Aircraft, 1976, p. 17/1-10, EEA80-26888.

1707. Greer, S. E. and Roush, W. B., "Modified bellows conductive cooling", IBM Tech. Disclosure Bull., 20 (11), 4391-2 (1978).

1708. Greeson, R. L. and Philofsky, E. M., "Metalization system for semiconductor devices", Patent USA 3985515, Publ. October 1976, CA85-185716.

1709. Gregor, L. V. and Hirsh, H., "Preventing hot-spot deposits in liquid cooled semiconductor device package", IBM Tech. Disclosure Bull., 19 (4), 1260 (1976).

1710. Gregor, L. V. and Reeber, M. D., "Articulated thermal conductor for semiconductor chip packages", IBM Tech. Disclosure Bull., 20 (8), 3131-2 (1978).

1711. Groh, L. H., "A shock-mounted, liquid cooled cold plate assembly", Report AD-D004774/6SL, Department of the Navy, Washington, D. C., 1977, 17 pp.

1712. Gupta, O. R., "Inner-fin air-cooled module", IBM Tech. Disclosure Bull., 19 (5), 1804 (1976).

1713. Gupta, O. R., "Adjustable cold plate chip cooling", IBM Tech. Disclosure Bull., 20 (3), 1120-1 (1977), EEA81-22843.

1714. Gupta, O. R. and Hwang, U. P., "Conduction cooling module with a variable resistance thermal coupler", IBM Tech. Disclosure Bull., 21 (6), 2434 (1978).

1715. Harpster, J. W. C., "Vacuum deposited TEDs for electronic device chip cooling", Proceedings of the Second International Conference on Thermoelectric Energy Conversion, 1978, p. 43-7, EEA81-49495.

1716. Hasson, J. C., Oktay, S. S. and Tomka, A. S., "Fluted pistons or piston walls for semiconductor chip cooling", IBM Tech. Disclosure Bull., 20 (6), 2219-20 (1977).

1717. Hirakawa, II., Oide, K. and Ogawa, Y., "Electronic cooling of resistors", J. Phys. Soc. Jpn., 44 (1), 337-8 (1978).

1718. Hosono, I. and Shikano, Y., "Vapor cooling device for semiconductor device", Patent USA 3989099, Publ. November 1976.

1719. Hu, S. and Mader, S. R., "Reducing chip fracture tendency by cushioning", IBM Tech. Disclosure Bull., 21 (7), 2747 (1978).

1720. Hultmark, E. B. and Pelletier, R. A., "Conduction cooling of semiconductor devices", IBM Tech. Disclosure Bull., 20 (11), 4819 (1978).

1721. Hultmark, E. B., Schiller, J. M. and Shepheard, R. G., "Thermal enhancement of modules", IBM Tech. Disclosure Bull., 19 (8) 3127 (1977).

1722. Huston, R. L., "Protective deflector for cooling fans", IBM Tech. Disclosure Bull., 21 (5), 1777 (1978).

1723. Hwang, U. P., "Conduction air-cooled module", IBM Tech. Disclosure Bull., 20 (3), 1007 (1977), EEA81-18372.
1724. Hwang, U. P., "Heat exchanger for vapour condensation by dropwise techniques", IBM Tech. Disclosure Bull., 20 (7), 2677 (1977), EEA81-45176.
1725. Hwang, U. P., Oktay, S., Pascuzzo, A. L. and Wong, A. C., "Conduction cooling module", IBM Tech. Disclosure Bull., 20 (11), 4334-5 (1978).
1726. Hwang, U. P. and Pascuzzo, A. L., "Liquid encapsulated conduction module", IBM Tech. Disclosure Bull., 20 (7), 2759-60 (1977), EEA81-45183.
1727. Hwang, U. P. and Spector, C. J., "Heat conductor module", IBM Tech. Disclosure Bull., 21 (1), 138 (1978).
1728. Ivanov, K. A., Janakiev, K. I. and Pozarlieva, E. A., "Radiators for semiconductor elements", Elektro Prom.-st. & Priborostr., 11 (6), 209-13 (1976), Bulgarian, EEA80-3476.
1729. Jeannotte, D. A., "Circuit module with gallium metal cooling structure", IBM Tech. Disclosure Bull., 19 (4), 1348 (1976).
1730. Johnson, A. H., "Diaphragm cooling for devices", IBM Tech. Disclosure Bull., 20 (8), 3121 (1978).
1731. Johnson, A. H., "Device cooling", IBM Tech. Disclosure Bull., 20 (10), 3919-20 (1978).
1732. Joshi, K. C., "Enhanced cooling of components on a card board", IBM Tech. Disclosure Bull., 20 (8), 3047-8 (1978).
1733. Kaiser, G. L., "Thermal management of the standard electronic module (SEM)", Proceedings of the IEEE 1977 National Aerospace and Electronics Conference, p. 86-93, EEA81-8094.
1734. Kammerer, H. C., "Thermal conduction button", IBM Tech. Disclosure Bull., 19 (12), 4622-3 (1977), EEA81-9251.
1735. Keller, C. J. and Moran, K. P., "Jet cooling cup for cooling semiconductor devices", IBM Tech. Disclosure Bull., 20 (9), 3575-6 (1978).
1736. Keyser, E. F. and Roedder, R., "A method for evaluating cooling requirements for high-speed circuits for high-density interconnections", 9th Annual Connector Symposium, 1976, p. 389-96, EEA80-13257.
1737. Koopman, N. G. and Totta, P. A., "Conduction cooled circuit package", Patent USA 4081825, Publ. March 1978.
1738. Krammer, H. C., "Thermal conduction button", IBM Tech. Disclosure Bull., 19 (12), 4622-23 (1977).
1739. Krogh, C., "Cooling of electronic equipment", Elektronik, no. 2, 32, 34 (1977), Danish, EEA81-101.
1740. Krumm, H., "Chip cooling", IBM Tech. Disclosure Bull., 20 (7), 2728-9 (1977), EEA81-45772.
1741. Larkin, B. S., "A computer cooling application for thermosiphons", Eng. J., 56 (1), 30-3 (1973), CCA8-11553.
1742. Leonard, I. M. and Axelband, S., "Cold plates and electronic packaging", Electron. Packag. & Prod., 17 (7), 209-12 (1977).

1743. Liander, W., "Encapsulation and cooling – important for component quality", Eltek. Aktuell Elektron., 21 (3), 20-2, 24 (1978), Swedish, EEA81-37132.

1744. Loeffel, E. G., Nutter, S. W. and Rzant, A. W., "Liquid cooled module with compliant membrane", IBM Tech. Disclosure Bull., 20 (2), 673-4 (1977), EEA81-18368.

1745. Lynch, J. R. and Walker, R. G., "Cooling assembly for integrated-circuit chip", IBM Tech. Disclosure Bull., 20 (1), 143 (1977), EEA81-14356.

1746. McDonald, J. A. and Weikel, W. J., "Custom tailoring of chip temperatures in a singular temperature environment", IBM Tech. Disclosure Bull., 21 (6), 2441-2 (1978).

1747. Markstein, H. W., "New developments in cooling techniques", Electron. Packag. & Prod., 16 (5), 36-8, 40, 44 (1975), EEA79-40776.

1748. Meeker, R. G., Scanlon, W. J. and Segal, Z., "Module thermal cap for semiconductor chip package", IBM Tech. Disclosure Bull., 20 (7), 2699-701 (1977), EEA81-45180.

1749. Metreaud, C. G., "Air-cooled semiconductor chip module configurations", IBM Tech. Disclosure Bull., 20 (7), 2697-8 (1977), EEA81-45179.

1750. Metreaud, C. G. and Schomaker, D. F., "Air-cooled semiconductor chip module configuration", IBM Tech. Disclosure Bull., 20 (7), 2695-6 (1977), EEA81-45178.

1751. Minchom, N. P., "Cooling of electronic equipment must be considered at the early design stage", Des. Eng., 43, 45, 47 (1975), EEA79-3725.

1752. Mohnot, S. R., Seetharamu, K. N. and Sastri, V. M. K., "Cooling of electronic equipment with base plate heat exchanger", Indian J. Technol., 13 (4), 156-9 (1975), EEA79-84.

1753. Moran, K. P., Pascuzzo, A. L. and Yacavonis, R. A., "Self-regulating integrated circuit module cold plate", IBM Tech. Disclosure Bull., 20 (9), 3439-42 (1978), EEA82-634.

1754. Moran, K. P., Pascuzzo, A. L. and Yacavonis, R. A., "Spring loaded module connectors for mounting an array of modules on circuit board", IBM Tech. Disclosure Bull., 20 (9), 3434-5 (1978).

1755. Ogiso, K. and Sasaki, E., "On cooling characteristics of a substrate with integrated circuits", Trans. Inst. Electron. & Commun. Eng. Jpn. Sect. E, E60 (1), 50 (1977), EEA80-41062.

1756. Oktay, S., "Bubble generation for cooling of semiconductor chips in liquid boiling regime", IBM Tech. Disclosure Bull., 19 (8), 3125-6 (1977), EEA80-31394.

1757. Oktay, S., "Substrate thermal plane", IBM Tech. Disclosure Bull., 21 (6), 2317 (1978).

1758. Oktay, S. and Pascuzzo, A. L., "Serrated pistons for improved module cooling", IBM Tech. Disclosure Bull., 21 (5), 1858-9 (1978).

1759. Pascuzzo, A. L. and Yacavonis, R. A., "Integrated circuit module package cooling structure", IBM Tech. Disclosure Bull., 20 (10), 3898-9 (1978).

1760. Pashley, R., Owen, W., Kokkonen, K. and Ebel, A., "Speedy RAM runs cool with power-down circuitry", Electronics, 50 (16), 103-7 (1977).

1761. Paul, G. T. and Wearmouth, J. W., "Optimization of cooling-water circuit design", Met. Technol., 5 (pt. 6), 203-11 (1978).

1762. Reeber, M. D., "Stabilizing liquid cooled semiconductor devices", IBM Tech. Disclosure Bull., 19 (4), 1261 (1976).

1763. Riseman, J., "Structure for cooling by nucleate boiling", IBM Tech. Disclosure Bull., 18 (77), 3700 (1976).

1764. Rockot, J., "Cooling methods for power semiconductors", Mach. Des., 50 (14), 93-4 (1978), EEA81-45185.

1765. Sachar, K. S., "Gas jet cooling with no net mechanical loading", IBM Tech. Disclosure Bull., 20 (9), 3723-4 (1978).

1766. Sachar, K. S., "Liquid jet cooling of integrated circuit chips", IBM Tech. Disclosure Bull., 20 (9), 3727-8 (1978), EEA82-630.

1767. Sedgwick, T. O., "Mounting technique for solder reflow areal array mounted silicon chips which allows direct and efficient backside heat removal", IBM Tech. Disclosure Bull., 20 (12), 5391 (1978).

1768. Shields, R. R., "Purifying per fluorocarbon coolants", IBM Tech. Disclosure Bull., 19 (4), 1264 (1976).

1769. Siegel, B. S., "Measuring thermal resistance is the key to a cool semiconductor", Electronics, 51 (14), 121-6 (1978), EEA81-41411.

1770. Simons, R. E. and Moran, K. P., "Immersion cooling systems for high density electronic packages", Proc. Tech. Program Natl. Electron. Packag. Prod. Conf., 1977, p. 396-409.

1771. Stuckert, P. E. and Toupin, R. A., "Distributed blowing engine for electronic cooling", IBM Tech. Disclosure Bull., 20 (4), 2445-6 (1977).

1772. Szabo, Z., "Dimensioning ventilation cooling on the basis of diagrams", Finommech. & Mikrotech., 15 (5), 142-6 (1976), Hungarian, EEA79-44755.

1773. Tadmor, M. and Frenkel, H., "Designing an optimum heat sink for forced-air cooling", Electron. Packag. & Prod., 17 (7), 258-62 (1977).

1774. Taylor, G., "Air cooling considerations in assembling electronics equipment (forced air cooling)", Circuits Manuf., 16 (7), 44-5 (1976), EEA79-40772.

1775. Vincent, G. A., "Dielectric liquid", Patent German 2608397, Publ. September 1976, CA85-152916.

1776. Vincent, G. A., "Dielectric liquid", Patent German 2608409, Publ. September 1976, CA85-152917.

1777. Vincent, G. A., "Dielectric liquid", Patent German 2608464, Publ. September 1976, CA85-152919.

1778. West, B., Coe, T. and Lavochkin, R., "Some considerations in clamping and cooling compression-type semiconductor devices", Conference Record of the IAS Annual Meeting, 1975, p. 35-42, EEA79-40777.

1779. Wilson, E. A., "True liquid cooling of computers", Proceedings of the 1977 National Computer Conference, AFIPS, Vol. 46, p. 341-8.

1780. Zimmerman, R., "Cooling devices for transistors", Radio Fersehen Elektron., 25 (22), 717-21 (1976), German, EEA80-20936.

1781. Zirnis, E. E., "Semiconductor module with improved air cooling", IBM Tech. Disclosure Bull., 20 (5), 1768 (1977), EEA81-32237.

## 2.   Heat Sinks/Heat Pipes

1782. Adamian, A., "Simple approach to cooling hot ICs with heat sinks", Mach. Des., 49 (12), 106-11 (1977).

1783. Andros, F. E. and Shott, F. A., "Flexible heat pipe", IBM Tech. Disclosure Bull., 20 (11), 4313-14 (1978).

1784. Anon., "'Thermal conduction pipe' and its applications", Elektronik, 24 (11), 101-2 (1975), German, EEA79-7832.

1785. Anon., "Thermal control of power supplies with electronic packaging techniques", Report N77-18386/1SL, Martin Marietta Corp., Denver, Colo., 1977, 145 pp.

1786. Antonetti, V. W., Hwang, U. P., Moran, K. P. and Pascuzzo, A. L., "Swivel piston conduction module", IBM Tech. Disclosure Bull., 20 (7), 2707-8 (1977), EEA80-45181.

1787. Basiulis, A. and Sekhon, K. S., "Heat pipes in electronic component packaging", Proc. Tech. Program Natl. Electron. Packag. Prod. Conf., 1977, p. 410-19.

1788. Bauman, A. H., Pittwood, D. G. and Weiss, R. L., "Multimodule heat sink", IBM Tech. Disclosure Bull., 19 (8), 2976-7 (1977), EEA80-35000.

1789. Birnbreier, H., Gross, F. and Heidtmann, U., "Heat pipe cooling of semiconductor power devices", 10th Intersociety Energy Conversion Engineering Conference, 1975, p. 1546-53, EEA79-44929.

1790. Blakeslee, A. E. and Hovel, H. J., "Photoelectrical converter", Patent USA 4062698, Publ. December 1977.

1791. Bond, G. L., Hawryluk, R. and Schmieg, J. W., "Wrap-around heat sink", IBM Tech. Disclosure Bull., 20 (4), 1434-5 (1977), EEA81-22847.

1792. Cechanek, R. and Joshi, K. C., "Module package", IBM Tech. Disclosure Bull., 21 (6), 2268-9 (1978).

1793. Clarke, S. C., "Heat sinking power semiconductors", Electron, no. 127, 27, 29, 31, 33 (1977), EEA81-23902.

1794. Colwell, G. T., "Prediction of cryogenic heat pipe performance", Report N77-76447/0SL, Georgia Inst. of Tech., Atlanta, 1977, 112 pp.

1795. Corso, J. A., Herdzik, R. J., Koopman, N. G. and Marcotte, V. C., "Solder reservoir for chip thermal connection", IBM Tech. Disclosure Bull., 20 (1), 142 (1977), EEA81-14355.
1796. Crosby, E. G., Fritz, H. E. and Hornbeck, F. C., "Heat sink", IBM Tech. Disclosure Bull., 19 (7), 2638 (1976), EEA80-31393.
1797. Dance, B., "Dissipating heat [from semiconductor devices]", Electron, no. 134, 40, 45 (1978), EEA81-42328.
1798. Dean, D. J., "Integral heat pipe package for microelectronic circuits", 2nd International Heat Pipe Conference, 1976, p. 481-502.
1799. DeMaine, F. J. and Hoffman, K. M., "Heat sink for electronic module", IBM Tech. Disclosure Bull., 20 (11), 4737 (1978).
1800. Doo, V. Y. and Johnson, A. H., "Controlled gap in semiconductor packages", IBM Tech. Disclosure Bull., 20 (4), 1433 (1977), EEA81-22846.
1801. Durand, R. D. and Hultmark, E. B., "Ceramic cap and heat sink for semiconductor package", IBM Tech. Disclosure Bull., 21 (3), 1064-5 (1978).
1802. English, D. L., Nakaji, E. M. and King, R. S., "Improved performance of millimetre-wave IMPATT diodes on type-IIa diamond heatsinks", Electron. Lett., 12 (25), 675-6 (1976), EEA80-9437.
1803. Ghosh, S., "Heat sink considerations for high power semiconductor devices", Electr. India, 16 (15), 29-35 (1976), EEA79-44925.
1804. Greenstein, G. M. and Simms, F., "Heat sinking wafer gripper", IBM Tech. Disclosure Bull., 20 (3), 949 (1977), EEA81-19392.
1805. Hardin, P. W., "Integral edge connector", IBM Tech. Disclosure Bull., 20 (11), 4346-8 (1978).
1806. Herdzik, R. J., Koopman, N. G. and Roush, W. B., "Externally applied chip thermal connection", IBM Tech. Disclosure Bull., 19 (9), 3368 (1977).
1807. Houslip, N., "Heat sinking for semiconductors", New Electron., 9 (12), 27, 30 (1976), EEA79-31897.
1808. Hudson, P. R. W., "The thermal resistivity of diamond heat-sink bond materials", J. Phys. D, 9 (2), 225-32 (1976), EEA79-12711.
1809. Hultmark, E. B., Miller, R. C. and Shepheard, R. G., "Dimensional control for semiconductor package", IBM Tech. Disclosure Bull., 21 (7), 2740-1 (1978).
1810. Hutchins, G. L., "Local deformation (mesa-approach) to diamond heat-sink imbedding", IBM Tech. Disclosure Bull., 20 (1), 427-8 (1977), EEA81-13293.
1811. Hutchins, G. L., "Batch-fabricated, polished copper heat-sink fabrication using an interconnecting web", IBM Tech. Disclosure Bull., 20 (1), 429-31 (1977), EEA81-13294.
1812. Hwang, U. P., "Heat exchanger for vapor condensation by drop-wise technqiue", IBM Tech. Disclosure Bull., 20 (7), 2677 (1977).

1813. Johnson, A. H., "Integrated circuit semiconductor device pack-
      age structure", IBM Tech. Disclosure Bull., 19 (9), 3387-8
      (1977), EEA80-36399.
1814. Johnson, A. H., "Edge connected chip carrier", IBM Tech. Dis-
      closure Bull., 21 (7), 2763-4 (1978).
1815. Keller, C. J. and Moran, K. P., "High power rectifier jet
      cooled heat sink", IBM Tech. Disclosure Bull., 21 (6), 2438
      (1978).
1816. Keller, C. J. and Moran, K. P., "High power rectifier evap-
      orative heat sink", IBM Tech. Disclosure Bull., 21 (6),
      2439-40 (1978).
1817. Kikuchi, K., Arisawa, K., Yuwata, Z. and Matsumoto, K., "Ap-
      plication of heat pipe", Furukawa Electr. Rev., no. 56, 67-80
      (1974), Japanese, EEA79-15887.
1818. Kilgenstein, O., "Problems of thermal conduction in elec-
      tronics. I.", Elektron. Ind., 6 (9), 174-7 (1975), German,
      EEA79-725.
1819. Koopman, N. G., "Extended chip package", IBM Tech. Disclosure
      Bull., 19 (9), 3369 (1977).
1820. Koopman, N. G., "Method for making conduction-cooled circuit
      package", Patent USA 4034468, Publ. July 1977; also, Patent
      USA 4034469, Publ. July 1977.
1821. Koopman, N. G. and Roush, W. B., "Solid state expanding alloys
      for chip thermal connection", IBM Tech. Disclosure Bull., 19
      (9), 3367 (1977).
1822. Kroeger, E., Morris, J., Miskolczy, G., Lieb, D. P. and Good-
      ale, D. B., "Performance of a thermionic converter module
      utilizing emitter and collector heat pipes", Report N78-
      27174/9SL, NASA, Lewis Research Center, Cleveland, Ohio,
      1978, 38 pp.
1823. Lindstron, C. and Jeppsson, B., "Heatsinking GaAs 4-18 GHz
      Gunn and IMPATT devices", Electron. Lett., 12 (3), 66-7
      (1976), EEA79-12907.
1824. Lipshutz, L. D. and Nutter, S. W., "Helium distribution scheme
      for thermal conduction modules", IBM Tech. Disclosure Bull.,
      20 (6), 2355-6 (1977), EEA82-108.
1825. McGroddy, J. C. and Zory, P. S., "Injection laser array",
      Patent USA 4069463, Publ. January 1978.
1826. Mehta, R. C., "Computer-aided optimum design of heat sinks",
      J. Inst. Electron. Telecommun. Eng., 24 (7), 298-301 (1978).
1827. Moran, K. P., Pascuzzo, A. L. and Yacavonis, R. A., "Spring
      loaded module connectors for mounting an array of modules
      on circuit board", IBM Tech. Disclosure Bull., 20 (9),
      3434-5 (1978).
1828. Nelson, L. A., Sekhon, K. S. and Fritz, J. E., "Improved MIC
      performance through internal heat pipe cooling", Proc. Tech.
      Program Natl. Electron. Packag. Prod. Conf., 1977, p. 441-7.
1829. Neumann, E. W., "Heat dissipation using thermally conducting
      putty", IBM Tech. Disclosure Bull., 19 (5), 1799 (1976).

1830. Neumann, E. W. and Rabenda, E. J., Jr., "Thermally conducting elastomeric device", Patent USA 4029999, Publ. June 1977.

1831. Nuccio, C. and Schiavone, R. N., "Circuit module with self-aligning heat transfer", IBM Tech. Disclosure Bull., 20 (6), 2329-30 (1977), EEA82-107.

1832. Ostergren, C. D., "Heat transfer chip", IBM Tech. Disclosure Bull., 20 (6), 2216 (1977).

1833. Patrickson, P., "Heat pipes. I. Design considerations and manufacture", Electr. Equip., 14 (10), 53, 55-6 (1975), EEA79-82.

1834. Pechenegov, Y. Y. and Serov, Y. I., "Minimization of the dimensions of heat tubes for cooling electronic components", Izv. VUZ Priborostr., 21 (1), 107-10 (1978), Russian, EEA81-32239.

1835. Rea, S. N. and West, S. E., "Thermal radiation from finned heat sinks", IEEE Trans. Parts, Hybrids & Packag., PHP-12 (2), 115-17 (1976), EEA79-36684.

1836. Rivenburgh, D. L. and Van Vestrout, V. D., "Backbonding an integrated circuit to a heat sink", IBM Tech. Disclosure Bull., 19 (10), 3692 (1977), EEA81-5641.

1837. Roberts, C. C., Jr., "Cooling arrays of circuit cards using heat pipes and forced air diffusers", Proceedings of the 11th Intersociety Energy Conversion Engineering Conference, 1976, p. 893-900, EEA80-16456.

1838. Ronkese, B. J., "Metal wool heat stud", IBM Tech. Disclosure Bull., 20 (3), 1122-3 (1977), EEA81-22844.

1839. Ronkese, B. J., "Centerless ceramic package with directly connected heat sink", IBM Tech. Disclosure Bull., 20 (9), 3577-8 (1978), EEA82-114.

1840. Ronkese, B. J., "Metal wool and indium heat sink", IBM Tech. Disclosure Bull., 21 (3), 1143-4 (1978).

1841. Scott, A. W., "Cool it: Hot semiconductors die before their time. Cool them off with the right heat sink/fan combination", Electron. Prod., 19 (7), 43-7 (1976).

1842. Shinoda, M., "Diamond heat sinks", Oyo Buturi, 44 (7), 789-94 (1975), Japanese, EEA79-7838.

1843. Shott, F. A., "Bow-tie heat sink", IBM Tech. Disclosure Bull., 20 (2), 563 (1977), EEA81-18366.

1844. Simons, R. E., "Detachable boiling heat sink", IBM Tech. Disclosure Bull., 21 (3), 1139-40 (1978).

1845. Smith, M. L., "DC power system", IBM Tech. Disclosure Bull., 20 (11), 4542-3 (1978).

1846. Swain, G., "How to design transistor heatsinks", Electron. Aust., 39 (8), 66-7, 69 (1977), EEA81-22841.

1847. Tadmore, M. and Frenkel, H., "Designing an optimum heat sink for forced-air cooling", Electron. Packag. & Prod., 17 (7), 258-62 (1977).

1848. Veloric, H. S., Presser, A. and Wozniak, F. J., "Ultra-thin RF silicon transistors with a copper-plated heat sink", RCA Rev., 36 (4), 731-43 (1975).

1849. Webbon, B. W., "Tubular sublimator/evaporator heat sink",
      Report N76-13599/5SL, NASA, Ames Research Center, Moffett
      Field, Calif., 1975, 12 pp.
1850. Wenthen, F., "Heat sink effect of printed conductors", IEEE
      Trans. Parts, Hybrids & Packag., PHP-12 (2), 110-15 (1976).
1851. Werner, K., "Heat-sinking crystals in ICs and individual s-c
      components", Funk-Tech., 31 (18), 560-70 (1976), German,
      EEA80-632.
1852. Yonezu, H., "Heat sink techniques for semiconductor devices",
      Oyo Buturi, 46 (2), 164 (1977), Japanese, EEA80-36205.

### 3.   Thermal Analysis

1853. Arnold, A. J., Clark, B. T., Dionne, E., Siwy, M. J., Jr. and
      Woodside, D. H., "Structural concept for module with improved
      heat transfer characteristics", IBM Tech. Disclosure Bull.,
      20 (7), 2675-8 (1977), EEA81-45175.
1854. Arnold, A. J., Clark, B. T., Dionne, E., Siwy, M. J. and Wood-
      side, D. H., "Electronic packaging structure", IBM Tech. Dis-
      closure Bull., 20 (11), 4820-2 (1978).
1855. Balderes, D. and Lynch, J. R., "Hot-pressed In-dot module",
      IBM Tech. Disclosure Bull., 20 (4), 1392-3 (1977), EEA81-22845.
1856. Balders, D., Lynch, J. R. and Yacavonis, R. A., "Heat dissi-
      pation from IC chips through module package", IBM Tech. Dis-
      closure Bull., 19 (11), 4165-6 (1977).
1857. Baxter, G. K., "Recommendation of thermal measurement tech-
      niques for IC chips and packages", 15th Annual Proceedings
      of Reliability Physics Symposium, 1977, p. 204-11.
1858. Baxter, G. K. and Anslow, J. W., "High temperature thermal
      characteristics of microelectronic packages", IEEE Trans.
      Parts, Hybrids & Packag., PHP-13 (4), 385-90 (1977),
      EEA81-13306.
1859. Baxter, G. K. and Lew, P. Y., "Thermal effect of material
      properties on electronic modules", Proc. Tech. Program Natl.
      Electron. Packag. Prod. Conf., 1977, p. 420-31.
1860. Bergles, A. E., Chu, R. C. and Seely, J. H., "Survey of heat
      transfer techniques applied to electronic packages", Proc.
      Tech. Program Natl. Electron. Packag. Prod. Conf., 1977,
      p. 370-85.
1861. Cammarano, A. S., Elliott, D. K. and Hinderer, J. J., "Chip
      temperature measurement during manufacture", IBM Tech. Dis-
      closure Bull., 21 (3), 1133 (1978).
1862. Cergel, L., "Thermal design considerations in CMOS circuits",
      Electron. Engineering, 48 (578), 56-7 (1976), EEA79-24899.
1863. Chu, R. C., "Conduction path for electronic components", IBM
      Tech. Disclosure Bull., 19 (11), 4279 (1977).
1864. Chu, R. C., "Internal thermal design of LSI package", IBM
      Tech. Disclosure Bull., 19 (12), 4690 (1977), EEA81-10373.

1865. Chu, R. C., "Design for providing thermal interface material between narrow thermal interface gaps", IBM Tech. Disclosure Bull., 20 (7), 2761-2 (1977).

1866. Chu, R. C. and Simons, R. E., "Thermal interface", IBM Tech. Disclosure Bull., 20 (2), 680-1 (1977), EEA81-18370.

1867. Coats, T., "Computer-aided thermal analysis for microcircuits", Electron. Packag. & Prod., 15 (10), 63-64, 67 (1975).

1868. Conigliari, P., "Simple heat equations for electronic circuits", Mach. Des., 50 (4), 104-7 (1978), EEA82-116.

1869. Cook, K. D., Jr., Kerns, D. V., Jr., Nagle, H. T., Jr., Slagh, T. D. and Ruwe, V. W., "Computer-aided thermal analysis of a hybrid multi-stage active bandpass filter/amplifier", IEEE Trans. Parts, Hybrids & Packag., PHP-12 (4), 344-50 (1976), EEA80-13120.

1870. David, R. F., "Computerized thermal analysis of hybrid circuits", IEEE Trans. Parts, Hybrids & Packag., PHP-13 (3), 283-90 (1977), EEA81-5635.

1871. Davis, C. E. and Coates, P. B., "Linearization of silicon junction characteristics for temperature measurement", J. Phys. E., 10, 613-16 (1977).

1872. Dean, D. J., "Thermal design aspects", In: Handb. Thick Film Technol., Holmes, P. J. and Loasby, R. G., (Eds.), Electrochem. Publications Ltd.: Ayr, Scotland, 1976, p. 228-57, CA88-97820.

1873. Dombroski, E. L., Hwang, U. P. and Pascuzzo, A. L., "Thermal conduction module", IBM Tech. Disclosure Bull., 20 (6), 2214-5 (1977), EEA82-105.

1874. Dryden, W. G., "Thick film thermal design", Circuits Manuf., 19 (1), 76, 78, 80 (1979).

1875. Dulnev, G. N. and Ermolina, E. I., "Thermal resistance of a system of parallelepipeds", J. Eng. Phys., 30 (5), 580-5 (1976), EEA80-41111.

1876. Ellison, G. N., "The thermal design of an LSI single-chip package", IEEE Trans. Parts, Hybrids & Packag., PHP-12 (4), 371-8 (1976), EEA80-12516.

1877. Ellison, G. N., "Theoretical calculation of the thermal resistance of a conducting and convecting surface", IEEE Trans. Parts, Hybrids & Packag., PHP-12 (13), 265-6 (1976), EEA80-46.

1878. Foster, R. A. and Spaight, R. N., "Thermally enhanced package for semiconductor devices", IBM Tech. Disclosure Bull., 20 (7), 2637-8 (1977), EEA81-45174.

1879. Frisch, D. and Ciccarone, R., "Thermal analysis for evaluating laminates", Circuits Manuf., 17 (7), 54-8 (1977), EEA81-19125.

1880. Hannemann, R., "Electronic system thermal design for reliability", IEEE Trans. Reliab., R-26 (5), 306-10 (1977), EEA81-18422.

1881. Hill, P. W. and Karr, P. C., "Thermal isolation technique for heat-sensitive components on a printed-circuit card", IBM Tech. Disclosure Bull., 20 (9), 3375-6 (1978), EEA82-611.

1882. Howard, R. T., "Semiconductor device package having a sub-
      strate with a coefficient of expansion matching silicon", IBM
      Tech. Disclosure Bull., 20 (7), 2849-50 (1977).
1883. Imre, L. and Danko, G., "Effect of casing design on temperature
      arise of encased electrical apparatus", Elektrotechnika, 69
      (6), 201-10 (1976), Hungarian, EEA79-44924.
1884. Kaiser, G. L., "Thermal management of the standard electronic
      module (SEM)", IEEE Proc. Natl. Aerosp. Electron. Conf.
      NAECON '77, 1977, p. 86-93.
1885. Kersuzan, G., "How to resolve thermal problems in hybrid
      microelectronics", Electron. & Microelectron. Ind., no. 218,
      24-9 (1976), EEA79-28705.
1886. Keyes, R. W., "Power dissipation in planar digital circuits",
      IEEE J. Solid-State Circuits, SC-10 (3), 181 (1975).
1887. Keyes, R. W., "A figure of merit for IC packaging", IEEE J.
      Solid-State Circuits, SC-13 (2), 265-6 (1978), EEA81-37133.
1888. Keyser, E. F. and Roedder, R., "A method for evaluating cool-
      ing requirements for high-speed circuits with high-density
      interconnections", 9th Annual Connector Symposium, 1976,
      p. 389-96, EEA80-13257.
1889. Kilgenstein, O., "Problems of heat dissipation in electronics.
      II.", Elektron. Ind., 6 (10), 196-7 (1975), German, EEA79-7833.
1890. Kinniment, D. J. and Edwards, D. A., "Thermal design in a
      hybrid system with high packing density", IEEE Trans. Com-
      ponents, Hybrids & Manuf. Technol., CHMT-1 (2), 176-81 (1978),
      EEA81-33027.
1891. Kobayashi, T., Speriosu, V. and Tocci, L. R., "Temperature
      distributions in a bubble chip", J. Appl. Phys., 49 (3),
      1936 (1978), EEA81-38458.
1892. Laff, R. A., Comerford, L. D., Crow, J. D. and Brady, M. J.,
      "Thermal performance and limitations of silicon-substrate
      packaged GaAs laser arrays", Appl. Opt., 17 (5), 778-84
      (1978), EEA81-29199.
1893. Leonard, I. M. and Axelband, S., "Cold plates and electronic
      packaging", Electron. Packag. & Prod., 17 (7), 209-12 (1977).
1894. Leonard, I. M. and Axelband, S. V., "Cold-plate thermal design,
      analysis and sizing", Electron. Packag. & Prod., 17 (10),
      101-2, 104-7 (1977).
1895. Leroy, Y., "The determination of temperature maps", Electron.
      & Appl. Ind., no. 236, 39-41 (1977), French, EEA81-5610.
1896. Maly, W. and Kramek, J. P., "Determination of the temperature
      distribution in a hybrid high-power device", Electron Tech-
      nol., 10 (4), 81-93 (1977), EEA81-19458.
1897. Martinot, H., Rossel, P., Vialaret, G. and Sarrabayrouse, G.,
      "A physical analysis of the operating temperature limits of
      complementary MOS transistor integrated circuits", 1st Eur-
      opean Solid State Circuits Conference-ESSCIRC, Extended
      Abstracts, 1975, p. 103, EEA79-28703.

1898. Mgebryan, R. G., "An analysis of the stationary thermal fields in uncased semiconductor devices with solid leads", Radio Eng. & Electron. Phys., 22 (4), 158-61 (1977), EEA81-33031.

1899. Minkin, S. B., Ulashchik, V. E., Fedorov, B. I. and Shashkov, A. G., "Heat transfer of thermistors in a nonuniform electric field", J. Eng. Phys., 29 (6), 1526-31 (1975), EEA80-28552.

1900. Newell, W. E., "Transient thermal analysis of solid-state power devices – making a dreaded process easy", IEEE Power Electronic Specialists Conference, 1975, p. 234-51, EEA79-46279.

1901. Nieberlein, V. A., "Thermal analysis of the Rfss switch driver", Report AD-A018170/1SL, Army Missile Research Development and Engineering Lab., Redstone Arsenal, Ala., 1975, 23 pp.

1902. Novak, V., "Heat transfer in integrated circuits", Acta Polytech. III, no. 1, 83-8 (1973), Czech, EEA79-20519.

1903. Nowak, S., "Thermal analysis and proper choice of interlayers plate in thick film circuits", Elektronika, 18 (6), 238-40 (1977), Polish, EEA80-35928.

1904. Oktay, S. S. and Pascuzzo, A. L., "Dendrite conduction module", IBM Tech. Disclosure Bull., 20 (6), 2218 (1977).

1905. Pierron, E. D. and Bobos, G. E., "Application of thermal analysis to integrated circuit molding compounds", J. Electron. Mater., 6 (3), 333-48 (1977), EEA81-19517.

1906. Pilkington, C. and Wadsworth, B., "Thermal design considerations for power devices", Electron. Eng., 49 (596), 91, 93-5, 97 (1977).

1907. Ritchie, K., "Application of thermal analysis of the semiconductor plastic package thermal shock testing reliability", Microelectron. & Reliab., 15 (5), 489-90 (1976), EEA80-5818.

1908. Ruwe, V. W. and Slagh, T. D., "Thermal analysis considerations of hybrid microelectronic circuits", Report AD-A022256/2SL, Army Missile Research Development and Engineering Lab., Redstone Arsenal, Ala., 1975, 68 pp.

1909. Sechi, F. N., Perlman, B. S. and Cusack, J. M., "Computer-controlled infrared microscope for thermal analysis of microwave transistors", Dig. Tech. Pap. IEEE MTTS Int. Microwave Symp., 1977, p. 143-6.

1910. Shott, F. A., "Boat thermal enhancement for semiconductor chips and modules", IBM Tech. Disclosure Bull., 20 (7), 2635-6 (1977), EEA81-45173.

1911. Siegal, B. S., "Use electrical tests for thermal measurements", Microwaves, 15 (6), 48, 57, 59 (1976).

1912. Spaight, R. N., "Thermal enhancement technique for large-scale integrated circuit chip carriers", IBM Tech. Disclosure Bull., 20 (7), 2614 (1977).

1913. Steinberg, D. S., "Quick way to predict temperature rise in electronic circuits", Mach. Des., 49 (2), 130-2 (1977).

1914. Steklov, V. K. and Kharzhevskii, R. A., "Thermal stabilisation of microcircuit substrates", Izv. VUZ Radioelektron., 20 (9), 112-15 (1977), Russian, EEA81-19514.

1915. Stewart, A., "Thermal considerations in digital circuit design", New Electron., 9 (20), 56, 58, 64, 68 (1976), EEA81-285.

1916. Suciu, P., Chisleag, R. and Cucurezeanu, I., "On obtaining thermal maps of semiconductor devices by hologram interfero-metry", Bul. Inst. Politeh. 'Gheroghe Gheorghiu-Dej' Bucuresti Ser. Electroteh., 39 (3), 11-16 (1977), Rumanian, EEA81-45961.

1917. Tomes, M., "The analysis of thermal fields in monolithic integrated circuits", Slaboproudy Obz., 37 (5), 216-22 (1976), Czech, EEA79-33137.

1918. Yu, E. Y., "Determination of temperature of integrated-circuit chips in hybrid packaging", IEEE Trans. Parts, Hybrids & Packag., PHP-12 (2), 139-46 (1976), EEA79-37713.

1919. Zilka, Z., "Calculation of the temperature field in a hybrid integrated circuit", Elektrotech. Cas., 29 (6), 427-36 (1978), Czech, EEA81-49501.

## XIX.   FABRICATION TECHNIQUES

1920. Anon., "Hybrid fabrication guidelines", Circuits Manuf., 16 (9), 22, 24-6 (1976), EEA80-9200.

1921. Aramati, V. S., Bitler, J. S., Pfahnl, A. and Shiflett, C. C., "Thin-film microwave integrated circuits", IEEE Trans. Parts, Hybrids & Packag., PHP-12 (4), 309-16 (1976), EEA80-13290.

1922. Aube, G., Beland, B., Koosis, A., Leroux, A. and Richard, S., "Thick film hybrid micro-electronics", Ingenieur, 62 (315), 3-8 (1976), French, EEA80-9202.

1923. Batev, K. P., Vacev, K. D. and Mateev, A. H., "Electrochemical tin-coating of terminal tape for thin film hybrid integrated circuits", Elektro Prom.-st. & Priborostr., 11 (2), 60-1 (1976), Bulgarian, EEA79-33156.

1924. Brauer, W., Kienel, G. and Wechsung, R., "Reproducible methods for the fabrication of microwave strip lines of high precision and reliability", Solid State Technol., 19 (12), 67-73 (1976), EEA80-13294.

1925. Carbaugh, J. S., Mayron, H. and Paul, R. J., "Retarding con-ductive paths between metal pads", IBM Tech. Disclosure Bull., 21 (7), 2917 (1978).

1926. Carroll, B. D., Cook, K. B., Kerns, D. V. and Slagh, T. D., "Development of technologies and procedures for advanced microcircuit prototype fabrication", Report AD-A058613/1SL, Auburn Univ., Ala., 1978, 268 pp.

1927. Cassagne, A., "The techniques of functional adjustment of hybrid circuits", Electron. & Microelectron. Ind., no. 209, 20-3 (1975), French, EEA79-9031.

1928. Clark, B. T. and Metreaud, C. G., "Integrated seal and signal band for stacked modules", IBM Tech. Disclosure Bull., 21 (6), 2318-19 (1978).

1929. Cometta, E., "Microelectronics in box assembly", Elettrificazione, no. 3, 139-41 (1976), Italian, EEA79-24876.

1930. Conta, R., "A consistent technique for assembling thin film hybrid circuits", Alta Freq., 45 (1), 72-7 (1976), Italian, EEA79-20555.

1931. Cross, R., Valentino, R. J. and Zabrenski, T. S., "Mechanism for centering substrates", Tech. Dig., no. 40, 7 (1975), EEA79-8977.

1932. Crowder, G. and McAllister, M., "Discrete wire shield frame for semiconductor chip packaging structure", IBM Tech. Disclosure Bull., 20 (1), 151 (1977), EEA81-14357.

1933. Curran, J. E., "Vacuum technologies applied to electronic component fabrication", J. Vac. Sci. & Technol., 14 (1), 108-13 (1977), EEA80-20884.

1934. Curran, J. E., Jeanes, R. V. and Sewell, H., "Technology of thin-film hybrid microwave circuits", IEEE Trans. Parts, Hybrids & Packag., PHP-12 (4), 304-9 (1976), EEA80-13289.

1935. Davies, R. and Newton, B. H., "Microwave hybrid integrated circuit technology", Wireless World, 84 (1506), 46-50, 67 (1978), EEA81-14346.

1936. Dittrich, F. J., Smyth, R. T. and Weir, J. D., "Thermal spraying. A new approach to thick film circuit manufacture", International Electrical, Electronics Conference and Exposition, 1977, p. 132-3, EEA81-23639.

1937. Doyle, K. B., "Some experience and conclusions using soldered and welded packages for hermetic thick film hybrids", Microelectron. & Reliab., 16 (4), 303-7 (1977), EEA81-10053.

1938. Duckworth, R. G., "High reliability Ta/Al thin-film hybrid circuits", Microelectron. & Reliab., 15 (2), 141-6 (1976), EEA79-33158.

1939. Duff, O. J., Koerckel, G. J., DeLuca, R. A., Mayer, E. H. and Worobey, W., "A high stability RC circuit using high nitrogen doped tantalum", Proceedings of the 28th Electronic Components Conference, p. 229-33, EEA81-37906.

1940. Ecker, M. E. and Olson, L. T., "Semiconductor package structure", IBM Tech. Disclosure Bull., 19 (11), 4167-8 (1977), EEA81-9249.

1941. Entschladen, H. and Nagel, U., "Microwave stripline techniques - a control system for microwave ICs. IV.", Elektron. Anz., 9 (8), 19-22 (1977), German, EEA81-14333.

1942. Freeman, G. S., "Substrate choice and resultant hybrid construction techniques", Microelectron. & Reliab., 14 (4), 371 (1975), EEA79-13055.

1943. Funari, J., Rivenburgh, D. L. and Turner, J. F., "High temperature solder mask [hybrid modules fabrication]", IBM Tech. Disclosure Bull., 20 (2), 547 (1977), EEA81-18350.

1944. Funk, W., "Thick-film technology", Philips Tech. Rev., 35 (5), 144-50 (1975), EEA79-8986.

1945. Gauthier, N., "Micro-circuits 'by the metre' on Capton film", Toute Electron., no. 420, 69-73 (1977), French, EEA81-10031.

1946. Green, W. and Albrecht, M. G., "Metal micropattern fabrication by the photolysis of lead iodide", Thin Solid Films, 37 (2), L57-60 (1976), EEA80-490.

1947. Griffiths, G. W., "Handling medium quantity production thick film hybrids", New Electron., 10 (6), 72, 74-5 (1972), EEA80-35934.

1948. Gruner, H., "Influence of sputtering conditions on metal films for thin-film circuits", NTG-Fachber., no. 60, 213-18 (1977), German, EEA81-23631.

1949. Guttridge, W. R., Hines, W. and Kehley, G. L., "Modularized PC assembly", IBM Tech. Disclosure Bull., 20 (6), 2162-3 (1977), EEA82-626.

1950. Haining, F. W. and Herbaugh, D. G., "Multilayer printed-circuit board", IBM Tech. Disclosure Bull., 20 (5), 1723 (1977), EEA81-37885.

1951. Hays, S. A., "Fabricating reliability into flexible printed wiring", Electron. Packag. & Prod., 17 (7), 111-17 (1977).

1952. Hetherington, D. R., "Construction and application of thick-film hybrid circuits", Elektron. Ind., 7 (5), 104-6 (1976), German, EEA79-37712.

1953. Heywood, D., "A heat resistant plastic substrate clad with sub-tractively processable resistive/conductive foil", NTG-Fachber., no. 60, 45-50 (1977), German, EEA81-23620.

1954. Hickman, H. H., Kalke, A. R., Lach, W. and Norman, L. J., "Hybrid integrated circuit assembly using computerized image analysis", Proceedings of the 28th Electronic Components Conference, 1978, p. 356-61, EEA81-42163.

1955. Hinuber, W., "Foil techniques - an alternative for the thin and thick film techniques at the manufacture of multichip-hybrid-modular units", Feingeraetechnik, 26 (6), 253-5 (1977), German, EEA80-35929.

1956. Hirsl, J., "New tendencies in the production of hybrid integrated circuits", Sdelovaci Tech., 25 (9), 335-6 (1977), Czech, EEA81-10056.

1957. Honn, J. J. and Stuby, K. P., "Electrical packages for LSI devices and assembly process therefor", Patent USA 4074342, Publ. February 1978.

1958. Horbert, T. A., Hultmark, E. B. and Shepheard, R. G., "Module stand-off concept", IBM Tech. Disclosure Bull., 20 (5), 1827 (1977), EEA81-37887.

1959. Johnson, D. R. and Uribe, F., "The effect of multiple refiring on the adhesion of thick-film conductors", Proceedings of the 1977 International Microelectronics Symposium, p. 1-7.

1960. Joly, J. and Ranger, J. B., "The use of magnetron sputtering for hybrid integrated circuits manufacturing", NTG-Fachber., no. 60, 15-19 (1977), German, EEA81-23641.

1961. Jourdain, P. and Petit, R., "Hybrid circuit manufacturing
      mechanization in R.T.C.", International Conference on Manu-
      facturing and Packaging Techniques for Hybrid Circuits, 1976,
      p. 99-106, French, EEA79-46561.
1962. Kaiser, B., "A new method for the manufacture of RC-thin
      film circuits using multilayer structures", NTG-Fachber.,
      no. 60. 207-12 (1977), German, EEA81-23630.
1963. Kaneko, Y., Iwasa, K. and Onishi, N., "One method of high-
      density assembly for LSI chips", Trans. Inst. Electron. and
      Commun. Eng. Jpn. Sect. E, E59 (7), 31-2 (1976), EEA80-41119.
1964. Katholing, G., "Highly integrated circuits - technology and
      application fields", Components Rep., 11 (4), 97-101 (1976),
      EEA80-6677.
1965. Khanna, S. P., "Hybrid integrated circuit including thick film
      resistors and thin film conductors and technique for fabrica-
      tion thereof", Patent USA 4031272, Publ. June 1977, CA87-61590.
1966. Korwin-Pawlowski, M., "Manufacturing methods and techniques
      for miniature high voltage hybrid multiplier modules", Report
      AD-A056011/OSL, Canadian Commercial Corp., Ottawa, 1978, 59 pp.
1967. Koszykowski, A. W., "Effective thick film production", Elec-
      tron. Prod. Methods & Equip., 4 (7), 48-9 (1975), EEA79-4733.
1968. Kruger, G., "Development and applications of thin film tanta-
      lum hybrid circuits", Funkschau, 48 (20), 853-6 (1976),
      German, EEA80-504.
1969. Kuhn, R., "High-frequency hybrid integration - the basis of
      progress in high-frequency apparatus techniques", Nachrich-
      tentech. Elektron., 25 (12), 445-51 (1975), German, EEA79-13056.
1970. Kun, L., "Hybrid realisation of high precision analog conver-
      sion modules", Electrocomponent Sci. & Technol., 5 (1),
      49-53 (1978).
1971. Kuntzleman, H. O., Legge, F. V. and Wasson, R. L., "Chip/
      cable field replaceable unit", IBM Tech. Disclosure Bull.,
      20 (7), 2619-20 (1977), EEA81-45771.
1972. Laczko, B. and Keri, I., "Assembly techniques in microelec-
      tronics", Finommech. & Mikrotech., 15 (3), 77-83 (1976),
      Hungarian, EEA79-33164.
1973. Lilen, H., "The micropresentation of integrated circuits by
      means of mounting them on film", Electron. & Microelectron.
      Ind., no. 212, 16-20 (1975), French, EEA79-13059.
1974. Lin, K. and Burden, J. D., "Mass spectrometric solutions to
      manufacturing problems in the semiconductor industry", J.
      Vac. Sci. & Technol., 15 (2), 373-6 (1978).
1975. Liu, C. N., "Matching the thermal coefficients of expansion
      of chips to module substrate", IBM Tech. Disclosure Bull.,
      19 (12), 4666-7 (1977), EEA81-10061.
1976. Lupinski, W., "Photo-chemical methods for IC manufacturing",
      Elektronika, 16 (11), 451-5 (1975), Polish, EEA79-13000.
1977. Lyman, J., "Film carriers win productivity prize (IC production,
      PCB production, wiring, techniques review)", Electronics, 48
      (21), 122-5 (1975), EEA79-4745.

1978. Lyman, J., "Special report: film carriers star in high-volume IC production", Electronics, 48 (26), 61-8 (1975), EEA79-9000.

1979. Martin, W., "Polymer thick film extend options for hybrid and PC fabrication", Circuits Manuf., 17 (5), 72, 75-7 (1977), EEA81-10042.

1980. Meinel, H., Rembold, B. and Wiesbeck, W., "Optimisation of thin and thick film technology for hybrid microwave circuits", Electrocompon. Sci. & Technol., 4 (3-4), 143-6 (1977), EEA81-14335.

1981. Miller, A. A. and Bouchard, J. G. F., "Automatic assembly of hybrid circuits on a ceramic substrate", Report AD-A035276/5SL, Harry Diamond Labs., Adelphi, Md., 1976, 19 pp.

1982. Miura, N., Fuura, Y., Kazami, A. and Yamagishi, M., "Insulated metal substrates for power hybrid ICs", Proceedings of the 1977 International Microelectronics Symposium, p. 223-7.

1983. Moulin, M. M. and Joly, M. J., "Co-sputtering of alloys used in hybrid microelectronics", Proc. Int. Vac. Congr., 7th, 1977, p. 1615-18, CA88-98034.

1984. **Narasimhan, T. R., "A novel hybrid packaging scheme for high** component density circuits", Proceedings of the 1977 International Microelectronics Symposium, p. 40-3.

1985. Nechaev, L. E., "System for control of parameters during production of integrated micro-circuits", Mekh. & Avtom. Proiz., no. 11, 25-7 (1975), Russian, EEA79-12990.

1986. Needham, V., "The production of complex thin film circuits", Electrocompon. Sci. & Technol., 3 (4), 209-16 (1977), EEA80-35904.

1987. Norwood, D., Laudel, A. and Blessner, P., "Manufacturing process for hybrid microcircuits containing (metallised) vias", IEEE Trans. Parts, Hybrids & Packag., PHP-12 (4), 323-35 (1976), EEA80-13292.

1988. Olesen, F. C., "Assembly of micropacks", Elektronik, 10, 30, 32 (1976), Danish, EEA80-3311.

1989. Patterson, F. K., Bacher, R. J., Mitchell, J. D. and Marcus, S. M., "The production of thick film multilayers using a copper/dielectric system", Elektron. Prax., 13 (4), 127-9, 131 (1978), German, EEA82-642.

1990. Pauza, W., "Dual relays in a DIP package", Proceedings of the 23rd Annual National Relay Conference, 1975, Paper 13, 6 pp.

1991. Peterson, R. E. and Baurle, K. L., "New metallization process for tantalum RC hybrid integrated circuits", 27th Electronic Components Conference, 1977, p. 238-44, EEA80-35948.

1992. Phillips, E., "Removing drill smear with plasma", Electron. Packag. & Prod., 17 (7), 138-43 (1977).

1993. Pichkhadze, I. P., Broisman, Z. G. and Nasidze, N. A., "Some automatic assemblies in integrated hybrid circuits", Prib. & Sist. Upr., no. 4, 44-6 (1976), Russian, EEA79-33161.

1994. Pops, H., "Manufacture of an integrated circuit package", Proceedings of the International Symposium on Shape-Memory Effects and Applications, Metallurgical Society of AIME, 1975, p. 525-36.

1995. RCA Solid State Technology Center, "Phase I final development report for high-reliability, low-cost integrated circuits", Report AD-A039954/3SL, RCA Solid State Technology Center, Somerville, N. J., 1977, 98 pp.

1996. Redemske, R. F., "Multilevel substrate technology and epoxy component attach for hybrid fabrication", 13th Annual Proceedings of Reliability Physics Symposium, 1975, p. 224-9, EEA79-46576.

1997. Redemske, R. F., "Fabrication technology for large scale hybrid microcircuits", 14th IEEE Computer Society International Conference, COMPCON 77, Digest of Papers, p. 253-7, EEA81-5638.

1998. Renz, H. W., "Fabrication and performance of highly miniaturized distributed RC-active circuits", NTG-Fachber., no. 60, 99-104 (1977), German, EEA81-23625.

1999. Rodrigues de Miranda, W. R. and Oswald, R. G., "Advances in TAB for hybrids - TC outer lead bonding", Proceedings of the 1977 International Microelectronics Symposium, p. 105-14.

2000. Roman, D. J. and Rutkiewski, J., "HYBRID integrated circuit processing", Tech. Dig., no. 45, 37-8 (1977), EEA80-35938.

2001. Ryan, P. M., "Registering multilayer ceramic substrates for E beam exposure", IBM Tech. Disclosure Bull., $\underline{20}$ (12), 5184-5 (1978).

2002. Samgouard, P. and Minault, M., "Hybrid thick film circuit technology use for airborne equipment", International Conference on Manufacturing and Packaging Techniques for Hybrid Circuits, 1976, p. 191-201, French, EEA79-46570.

2003. Sheets, L. R., "Input/output techniques for computerized hybrid-microcircuit masking", Microelectronics, $\underline{5}$ (3), 50-6 (1974), EEA79-9040.

2004. Siegle, G., "Hybram - hybrid-assembled module", Funkschau, $\underline{49}$ (5), 195-7 (1977), German, EEA80-35930.

2005. Smith, J. M. and Stuhlbarg, S. M., "Hybrid microcircuit tape chip carrier materials/processing trade-offs", IEEE Trans. Parts, Hybrids & Packag., PHP-$\underline{13}$ (3), 257-68 (1977).

2006. Snyder, H. B., "Controlling over-etch during PCB manufacturing", Circuits Manuf., $\underline{17}$ (7), 28-34 (1977).

2007. Sun, Y. E. and Driscoll, J. C., "Direct bonded, glass passivated power modules: the new 'IC's' of power electronics", Conference Record of the IAS Annual Meeting, 1975, p. 50-4, EEA79-37738.

2008. Teredesai, M. P., Ajit, C. N. and Jawalekar, S. R., "Thin-film hybrid circuits", Int. J. Electron., $\underline{44}$ (1), 41-8 (1978), EEA81-14319.

2009. Traeger, R. K., "Considerations in lid sealant for hybrid microelectronic circuits", Electrochemical Society Spring Meeting, Extended Abstracts, 1975, p. 305-16, EEA79-1110.

2010. Van Nie, A. G., "Electroless NiP processing for hybrid integrated circuits", Microelectron. & Reliab., $\underline{15}$ (3), 221-6 (1976), EEA79-37715.

2011. Van Vestrout, V. D., "Hybrid module including thin-film con-
      ductors and paste resistors", IBM Tech. Disclosure Bull., 19
      (12), 4541 (1977), EEA81-10050.
2012. Veselov, L. V. and Mazur, A. I., "Aspects of automated assem-
      bly in microelectronics", Izv. VUZ Priborostr., 18 (10),
      115-21 (1975), Russian, EEA79-13061.
2013. Volkova, I. A., Grishchuk, S. A., Dmitrenko, E. I. and Seni-
      shin, Y. M., "Technological aspects of manufacture of hybrid
      integrated circuits at microwaves", Poluprovodn. Tekh. &
      Mikroelektron., no. 24, 67-71 (1976), Russian, EEA80-508.
2014. Walsby, A. M., "Off the buses", Systems, 4 (5), 24-5 (1976),
      EEA79-46580.
2015. Wellby, J. P., "Development of film hybrid circuits", New
      Electron., 10 (6), 79-80 (1972), EEA80-35935.
2016. White, C. E. T. and Sohl, H. C., "Proforma on preforms",
      Solid State Technol., 45-8 (1975), EEA79-730.
2017. Yamamoto, N., "Manufacturing process of a laminated type
      beryllium oxide package for electronic use", Process. Kinet.
      Prop. Electron. Magn. Ceram. Proc. U.S.-Jpn. Semin. Basic Sci.
      Ceram., 1976, p. 17-23, CA88-157285.

XX.   REPAIR/REWORK

2018. Anon., "The repair or modification of pads, tracks and plated-
      thru holes on PCBs", Electron. Prod. Methods & Equip., 6 (3),
      19, 21 (1977), EEA81-14261.
2019. Fancher, D. R. and McDaniel, J. E., "Seal and rework evalua-
      tion of seam-welded hybrid packages", Electron. Packag. &
      Prod., 18 (2), 88-90, 92, 94 (1978), EEA82-104.
2020. Funari, J., Jumbeck, G., Moore, R. J. and Myers, F. R., "Braz-
      ing for module pin repair", IBM Tech. Disclosure Bull., 20
      (11), 4307-8 (1978).
2021. Hannan, J. J., Shapiro, M. R. and Walker, W. J., "Reworkable
      cover plate for semiconductor device packages", IBM Tech. Dis-
      closure Bull., 21 (1), 145 (1978).
2022. Ho, C. W. and Ting, C. Y., "Planar wiring repair technique
      for the tin-film metal package via solder evaporation", IBM
      Tech. Disclosure Bull., 20 (9), 3729-30 (1978), EEA82-115.
2023. Johnson, C., Jr., Morrissey, J. M. and Scrivner, C. H., "Re-
      moval technique [for integrated circuit chip pads]", IBM
      Tech. Disclosure Bull., 20 (6), 2209-10 (1977), EEA82-632.
2024. Khalil, Z., "On the reliability of a 3-unit redundant system
      with two types of repair", Microelectron. & Reliab., 16 (6),
      689-90 (1977), EEA81-9284.
2025. Kontoleon, J. M., "Optimum redundancy of repairable modules",
      IEEE Trans. Reliab., R-26 (4), 277-9 (1977).
2026. Locke, M. C., Scardino, W. and Croop, H., "Non-tank phosphoric
      acid anodize method of surface preparation of aluminum for
      repair bonding", National SAMPE Technical Conference Vol. 7,
      1975, p. 488-504.

2027. Mathers, M. R. and Muller, J. B., "Chip removal and substrate rework tool", IBM Tech. Disclosure Bull., <u>21</u> (3), 1216–17 (1978).

2028. Plesser, K. T. and Field, T. O., "Cost-optimized burn-in duration for repairable electronic systems", IEEE Trans. Reliab., R-<u>26</u> (3), 195–7 (1977), EEA80-40381.

2029. Pottier, B., "Method of reworking sealed modules", IBM Tech. Disclosure Bull., <u>20</u> (2), 624 (1977), EEA81-14359.

2030. Rubinstein, M., "Precision circuit repair with selective plating", Insul./Circuits, <u>23</u> (8), 37–40 (1977), EEA81-903.

2031. Sachar, K. S. and Sedgwick, T. A., "Use of a heated gas jet to remove a silicon chip soldered to a substrate", IBM Tech. Disclosure Bull., <u>20</u> (9), 3725–6 (1978), EEA82-652.

2032. Wallgren, L. and Rosenthal, A., "Circuit board repairs with parts kit and soldering iron", Insul./Circuits, <u>23</u> (8), 15–17 (1977), EEA81-902.

2033. Walther, V., Horn, J. and Neukirchner, W., "Soldering repair point RLM1", Radio Fernsehen Elektron., <u>26</u> (21–22), 741 (1977), German, EEA81-33004.

## XXI.  TYPES OF PACKAGES
### 1.  Bubble Domain

2034. Anon., "Open coil structure for bubble-memory device packaging", Comput. Des., <u>15</u> (9), 112 (1976), CCA12-3992.

2035. Anon., "Bubble-memory packaging alternatives: producibility vs. interchangeability", Digital Des., <u>7</u> (1), 54–5 (1977), EEA80-21112.

2036. Becker, F. J. and Stermer, R. L., "Packaging of a large capacity magnetic bubble domain spacecraft recorder", 14th IEEE Computer Society International Conference, Digest of Papers, 1977, p. 258-62.

2037. Bobeck, A. H. and Danylchuk, I., "Characterization and test results for a 272K bubble memory package", IEEE Trans. Magn., MAG-<u>13</u> (5), 1370–2 (1977), EEA80-41811.

2038. Braun, R. J., "Rotational field generator for bubble memory package", Patent USA 4090252, Publ. May 1978.

2039. Braun, R. J. and Burry, J. M., "Coil design for bubble memory package", IBM Tech. Disclosure Bull., <u>19</u> (9), 3323–4 (1977), EEA80-32591.

2040. Braun, R. J. and McBride, D. G., "Double flat coil for bubble memory package", IBM Tech. Disclosure Bull., <u>19</u> (10), 3713–14 (1977), EEA81-6242.

2041. Brumley, R. L., "Bubble memory chip mounting", IBM Tech. Disclosure Bull., <u>16</u> (10), 3128–9 (1974), CCA9-18394.

2042. Carden, G. R. and McBride, D. G., "Insulator system", IBM Tech. Disclosure Bull., <u>20</u> (11), 4753 (1978).

2043. Carden, G. R., McBride, D. G. and Spaight, R. N., "Flexible bubble memory package", IBM Tech. Disclosure Bull., <u>21</u> (6), 2285–6 (1978).

2044. Carlo, J. T., Stephenson, A. D., Jr. and Hayes, D. J., "Chip packing efficiency in bubble memories", IEEE Trans. Magn., MAG-12 (6), 624-8 (1976), CCA12-3995.

2045. Chen, T. T. and Bohning, O. D., "Investigation of system integration methods for bubble domain flight recorders", Report N75-26336/8ST, Rockwell International Corp., Anaheim, Calif., 1975, 240 pp.

2046. Chen, T. T., Oeffinger, T. R. and Gergis, I. S., "A hybrid decoder bubble memory organization", IEEE Trans. Magn., MAG-12 (6), 630-2 (1976), EEA80-6831.

2047. DeBonte, W. J. and Butherus, A. D., "Magnetically permeable adhesives and adhesive-jointed shield structures", IEEE Trans. Magn., MAG-13 (5), 1376-8 (1977).

2048. DeBonte, W. J. and Zappulla, R., "Relationship of bias field setting procedure to field stability against external field perturbations for magnetic bubble memory bias field structures", IEEE Trans. Magn., MAG-12 (6), 645-7 (1976).

2049. Gibson, A. T. and Hunter, D. J., "Magnetic bubble memory system design and evaluation using single chip packages", 14th IEEE Computer Society International Conference, Digest of Papers, 1977, p. 236-8, EEA81-1498.

2050. Hagedorn, F. B., Rago, L. F., Kish, D. E., Chen, Y. S., Hess, W. E., Beurrier, H. R. and Wagner, W. D. P., "Magnetic bubble device testing", IEEE Trans. Magn., MAG-13 (5), 1364-9 (1977).

2051. Kiseda, J. R., "High speed stripline coils for magnetic bubble devices", IEEE Trans. Magn., MAG-9 (3), 425-9 (1973).

2052. Kotrch, G. S. and Miesch, E. D., "Wire guide for coil winding machine", IBM Tech. Disclosure Bull., 20 (8), 3032-3 (1978).

2053. Kowalchuk, R., Bobeck, A. H., Butherus, A. D., Ciak, F. J. and Smith, D. H., "Magnetic bubble dual-in-line package (DIP) functional and reliability testing", IEEE Trans. Magn., MAG-12 (6), 691-3 (1976), EEA80-6848.

2054. Larnerd, J. D., "Bubble domain package", IBM Tech. Disclosure Bull., 21 (7), 2714-15 (1978).

2055. Lyons, W. A., "Permanent magnet bias schemes for bubble memory applications", 1973 Digest of the Intermag. Conference, Pap. 26-6, 1 pp.

2056. McBride, D. G., "Double stacked bubble memory module", IBM Tech. Disclosure Bull., 20 (8), 3054-5 (1978).

2057. Maegawa, H., Matsuda, J. and Takasu, M., "Flat packaging of magnetic bubble devices", IEEE Trans. Magn., MAG-10 (5), 753-6 (1974).

2058. Naden, R. A., Keenan, W. R. and Lee, D. M., "Electrical characterization of a packaged 100K bit major/minor loop bubble device", IEEE Trans. Magn., MAG-12 (6), 685-7 (1976), EEA80-6846.

2059. Rifkin, A. A., "A practical approach to packaging magnetic bubble devices", IEEE Trans. Magn., MAG-9 (3), 429-33 (1973), EEA77-6244.

2060. Rifkin, A. A., "Non volatile magnetic domain device having binary value bias field excitation", Patent USA 3836896, Publ. September 1974.

2061. Rifkin, A. A., Bagholtz, W. B., Bosch, L. S., Downing, R. A., Kiseda, J. R., Lennon, A. L., Jr. and Scott, E. W., "Packaged magnetic domain device having integral bias and switching magnetic field means", Patent USA 3958155, Publ. May 1976.

2062. Takasu, M., Maegawa, H., Furuichi, S., Okada, M. and Yama-gishi, K., "Fast access memory design using 3 μm bubble 80K chip", IEEE Trans. Magn., MAG-12 (6), 633-5 (1976).

2063. Takasu, M., Maegawa, H., Sukeda, T. and Yamagishi, K., "Re-flection coil packaging for bubble devices", IEEE Trans. Magn., MAG-11 (5), 1151-3 (1975), EEA79-4981.

2064. Uberbacher, E. C., "Bubble memory package via eggerate", IBM Tech. Disclosure Bull., 14 (11), 3378 (1972), EEA75-28072.

2065. Ypma, J. E. and Swanson, P., "Design and performance of a 100K byte serial bubble recorder", Dig. Pap. IEEE Comput. Soc. Int. Conf., 14th, 1977, p. 239-42.

2066. Yu, E. Y., "Thermal characteristics of a four-chip magnetic bubble package", IEEE Trans. Magn., MAG-13 (5), 1373-5 (1977), EEA80-41812.

## 2.  Ceramic

2067. Anazawa, S. and Kuroda, T., "Ceramic substrate assembly for electronic circuits having ceramic films thereon for inter-cepting the flow of brazing agents", Patent USA 3908184, Publ. September 1975.

2068. Anon., "Low cost substrate for thick film circuits using porcelain", New Electron., 11 (3), 36 (1978), EEA81-23633.

2069. Fleischner, P. L., "Beryllia ceramics in microelectronic applications", Solid State Technol., 20 (1), 25-30 (1977).

2070. Gardner, R. A., "High-alumina content compositions containing BaO-MgO-SiO$_2$ glass and sintered ceramic articles made there-from", Patent USA 4020234, Publ. April 1977.

2071. Handschuh, K., Luft, H., Roske, E. and Zeibig, A., "Ceramic materials in electronics", Funkschau, 48 (19), 805-8 (1976), German, EEA79-46584.

2072. Kaiser, H. D. and Nufer, R. W., "Ceramic substrate", Patent USA 3899554, Publ. August 1975.

2073. Kaneko, Y., Iwasa, K. and Onishi, N., "One method of high-density assembly for LSI chips", Trans. Inst. Electron. & Commun. Eng. Jpn. Sect. E, E59 (7), 31-2 (1976), EEA80-41119.

2074. Niwa, K., Nakamura, J., Murakawa, K. and Nakamura, M., "New alumina substrate for hybrid integrated circuits", Proceed-ings of Electron. Components Conference, Vol. 24, 1974, p. 105-10, CA85-12883.

2075. Peters, R. D., "Ceramic flat pack enclosures for precision quartz crystal units", Proceedings of the 30th Annual Sympos-ium on Frequency Control, 1976, p. 224-31, EEA80-17314.

2076. Smyly, J. P., "Growth of thick and film applications fosters ceramic substrate advances", Insul./Circuits, 24 (4), 49-51 (1978), EEA81-42151.
2077. Thomas, R. N., "Insulator substrate with a thin monocrystalline semiconductive layer", Patent USA 3902979, Publ. September 1975.
2078. Williams, J. C., "Evolution of ceramic substrates for thick and thin film components and circuits in the U.S.", Am. Ceram. Soc. Bull., 56 (6), 580-5 (1977).

### 3.  Hybrid

2079. Anderson, J. and Santana, J., "Specify hybrid components properly", Electron. Des., 24 (3), 52-7 (1976), EEA79-20560.
2080. Anon., "Hybrid thick-layer technique in micro electronics", Polytech. Tijdschr. Elektrotech. Elektron., 32 (4), 201-4 (1977), Dutch, EEA80-24623.
2081. Anon., "Mini-packaging for hybrid circuits", Electron. & Appl. Ind., no. 250, 67 (1978), French, EEA81-45763.
2082. Aube, G., Beland, B., Kocsis, A., Leroux, A. and Richard, S., "Thick film hybrid micro-electronics", Ingenieur, 62 (315), 3-8 (1976), French, EEA80-9202.
2083. Berger, R. L., Fabac, S. M., Jr., Gassner, G. E. and Wolz, G. N., "Circuit package", Tech. Dig., no. 46, 3 (1977), EEA81-14344.
2084. Besamat, M., "Hybrid circuit overtaking features", International Conference on Manufacturing and Packaging Techniques for Hybrid Circuits, 1976, p. 173-80, French, EEA79-46568.
2085. Boswell, D., "Customer-manufacturer relationship in hybrid microelectronics", International Conference on Manufacturing and Packaging Techniques for Hybrid Circuits, 1976, p. 35-46, EEA79-46559.
2086. Bouin, J., "CIT-ALCATEL's hybridation prospects in connection with technical, technological and economical criteria", International Conference on Manufacturing and Packaging Techniques for Hybrid Circuits, 1976, p. 47-56, French, EEA79-46560.
2087. Bouvier, J. and De Keyser, A., "Network thick-film printed networks for hybrid circuit applications", International Conference on Manufacturing and Packaging Techniques for Hybrid Circuits, 1976, p. 57-78, French, EEA79-46397.
2088. Brodersen, A. J. and Kinser, D. L., "Hybrid microelectronics at Vanderbilt University", Proceedings of the 1976 IEEE Southeastcon Region 3 Conference on Engineering in a Changing Economy, p. 275-7, EEA79-37737.
2089. Cagle, R. H., "Variable frequency clock driver: circuit hybridization", Proceedings of the 1977 International Microelectronics Symposium, p. 300-2.
2090. Calkins, R. and Berg, A., Jr., "Data acquisition in a DIP shrinks systems", Electronics, 49 (14), 77-83 (1976), EEA79-37709.

2091. Carpanini, E., "Thin film hybrids: a brief look at the poor
      man's LSI", Electron, no. 103, 58, 60 (1976), EEA80-502.
2092. Castracasi, G. and Gorla, C., "Investigations into production
      standards for thick film hybrid circuits", Alta Freq., 45 (1),
      77-81 (1976), Italian, EEA79-20556.
2093. Coles, J., "The role of the hybrid microcircuit in professional
      electronics", New Electron., 10 (6), 66, 69, 71 (1972),
      EEA80-35933.
2094. de Kock, D. L., "Thick film hybrids in New Zealand", N. Z.
      Eng., 31 (6), 162-4 (1976), EEA79-46396.
2095. De LaMoneda, F. H., Debar, D. E., Stuby, K. P. and Bertin,
      C. L., "Hybrid protective device for MOS-LSI chips", IEEE
      Trans. Parts, Hybrids & Packag., PHP-12 (13), 172-5 (1976),
      EEA80-3541.
2096. Dodd, H., "Hybrid thick films", Electron, no. 137, 17 (1978),
      EEA81-49502.
2097. Easson, R. M., "Thin film hybrid microcircuits below 500 MHz",
      New Electron., 10 (17), 18, 21-2 (1977), EEA81-14320.
2098. Epand, D., "Use SOT-23 packaged components in mass-produced
      hybrid circuits", Electron. Des., 24 (16), 66-70 (1976),
      EEA79-46545.
2099. Evans, J., "Hybrid devices (ICs)", Electron. Ind., 1 (3),
      25-7, 29 (1975), EEA79-4781.
2100. Gillis, T. B. and Ciccio, J., "The future of hybrids in an
      MSI/LSI world", Proceedings of the 1976 IEEE Southeastcon
      Region 3 Conference on Engineering in a Changing Economy,
      p. 214-6, EEA79-37735.
2101. Gotra, Z. Y., Dzisyak, E. P., Kuzkin, A. P., Radukhovskii,
      F. I. and Smelyanskii, I. L., "Private use hybrid integrated
      circuits for instrument construction", Prib. & Sist. Upr.,
      no. 4, 23-4 (1976), Russian, EEA79-33160.
2102. Hetherington, D. R., "Thick-film hybrid microcircuits", Mundo
      Electron., no. 58, 63-6 (1976), Spanish, EEA80-13293.
2103. Hicks, R. E. and Zimmerman, D. D., "Thin film hybrid micro-
      circuits on polymer substrates", Proceedings of the 1976 IEEE
      Southeastcon Region 3 Conference on Engineering in a Changing
      Economy, p. 217-8, EEA79-37736.
2104. Himmel, R. P., "A new generation of large hybrid modules -
      SLIM", Solid State Technol., 20 (10), 68-73 (1977), EEA81-19167.
2105. Isert, H., "Application of hybrid modules in telecommunica-
      tions", Nachr. Elektron., 30 (11), 253-6 (1976), German,
      EEA80-7380.
2106. Jackson, C. R., "Inexpensive covers for large microwave hy-
      brids", 12th Electrical/Electronics Insulation Conference,
      1975, p. 108-10, EEA79-46556.
2107. Jan, F., Hrovat, M. and Kolar, D., "Passive and active thick
      film circuits for professional and commercial electronics",
      Elektroteh. Vestn., 44 (3), 157-62 (1977), Slovene, EEA81-14322.

2108. Kirby, P. L., "The application of improved thick film techniques and materials in the production of complex hybrid
modules", International Conference on Manufacturing and Packaging Techniques for Hybrid Circuits, 1976, p. 155-62,
EEA79-46566.

2109. Kruger, G., "Development and applications of thin film tantalum hybrid circuits", Funkschau, 48 (20), 853-6 (1976),
German, EEA80-504.

2110. Lambert, F., "Packages and terminals: selection of the correct
metal package for a hybrid circuit", Electron. & Microelectron. Ind., no. 209, 58-61 (1975), EEA79-7834.

2111. Lemeunier, P., "Microelectronic hybrids: what stage are we
now?", Electron. & Appl. Ind., no. 231, 17-22 (1977), French,
EEA80-28628.

2112. Lemeunier, P., "New applications of thick film technology to
hybrids", Electron. & Appl. Ind., no. 238, 27-9 (1977),
French, EEA81-10047.

2113. Markstein, H. W., "Developments in substrates and packages
for hybrid circuits", Electron. Packag. & Prod., 18 (9),
50-1, 52, 54 (1978).

2114. Marshall, W. E., "Single hybrid package houses 12-bit data-
acquisition system", Electronics, 51 (13), 113-18 (1978),
EEA81-42158.

2115. Muccino, F. R., "A unique hybrid packaging concept for military applications", 8th Annual Connector Symposium, 1975,
p. 227-36, EEA79-3734.

2116. Murphy, C., "Hybrid technology loose particles and coating
materials", 13th Annual Proceedings of Reliability Physics
Symposium, 1976, p. 248-52, EEA79-46579.

2117. Okoshi, T. and Takeuchi, T., "Planar 3-dB hybrid circuit",
Electron. & Commun. Jpn., 58 (8), 80-90 (1975), EEA80-20583.

2118. Palma, F. and Scatamacchia, A., "Microcomponents for hybrid
integrated microwave circuits",, Alta Freq., 45 (1), 81-5
(1976), Italian, EEA79-20557.

2119. Palmer, D. W., "Hybrid microcircuitry for 300°C operation",
IEEE Trans. Parts, Hybrids & Packag., PHP-13 (3), 252-7 (1977).

2120. Pantanelli, G. P. and Sedora, E. J., "The selection of conductor materials for thick film hybrid circuits", International Conference on Manufacturing and Packaging Techniques
for Hybrid Circuits, 1976, p. 133-40, EEA79-46564.

2121. Peterson, R. E. and Baurle, K. L., "New metallization process
for tantalum RC hybrid integrated circuits", 27th Electronic
Components Conference, 1977, p. 238-44, EEA80-35948.

2122. Sergent, J. E., "Recent advances in thick film hybrids",
International Microelectronics Conference, Proceedings of the
Technical Program, 1975, p. 167-71.

2123. Startin, R. A. and Cross, A. C., "A hybrid thin-film logic
circuit using gallium arsenide field effect transistors",
Electrocompon. Sci. & Technol., 5 (2), 113-17 (1978),
EEA81-42160.

2124. Sutton, L., "Semiconductor memories packaged in hybrid micro-
      circuits", 27th Electronic Components Conference, 1977,
      p. 10-15, EEA80-35943.
2125. Taylor, J. L. and Prigel, D., "(Thin-film microstrip) wiggly
      phase shifters and directional couplers for radio-frequency
      hybrid-microcircuit applications", IEEE Trans. Parts, Hybrids
      & Packag., PHP-12 (4), 317-23 (1976), EEA80-13291.
2126. Tiesnes, M., "Monolithic integration and hybrid technology",
      Tech. Rundsch., 70 (17), 37, 39 (1978), German, EEA81-37921.
2127. Weiland, D., "Hybrid circuits shed the 'exotic' label", Mach.
      Des., 50 (14), 74-9 (1978), EEA81-45765.
2128. Whitehead, J., "Thick film circuits - what must be done to
      meet the demand", Electron. Equip. News, 4-5 (1977), EEA80-17041.
2129. Whitelaw, D., "Applications of thick film hybrids", New
      Electron., 9 (6), 66, 69, 71 (1976), EEA79-28707.
2130. Whitelaw, D., "Thick film hybrid microcircuits - general ap-
      plications (reasons to use them)", Microelectron. & Reliab.,
      15 (4), 335-8 (1976), EEA79-46549.
2131. Zuch, E. L., "Hybrid and monolithic data conversion circuits",
      New Electron., 10 (22), 50-62 (1977).

## 4. Microwave

2132. Anazawa, S., "Packaged semiconductor device for microwave
      use", Patent USA 3908186, Publ. September 1975, EEA79-8912.
2133. Anon., "Microwave stripline technology - a conductor system
      for integrated microwave circuits. II.", Elektron. Anz., 9
      (5), 34-7 (1977), German, EEA81-10043.
2134. Aramati, V. S., Bitler, J. S., Pfahnl, A. and Shiflett, C. C.,
      "Thin-film microwave integrated circuits", IEEE Trans. Parts,
      Hybrids & Packag., PHP-12 (4), 309-16 (1976), CA86-99860.
2135. Caton, W. A., III, Sergent, J. E. and Allen, J. L., "Micro-
      wave filters using microelectronic conductor deposition tech-
      nology", Proceedings of the 1977 International Microelectronics
      Symposium, p. 181-4.
2136. Davies, R. and Newton, B. H., "Microwave hybrid integrated
      circuit technology", Wireless World, 84 (1506), 46-50, 67
      (1978), EEA81-14346.
2137. Entschladen, H. and Nagel, U., "Microwave stripline technology
      - a guidance system for integrated microwave circuits. II.",
      Elektron. Anz., 9 (6), 20-2 (1977), German, EEA81-10044.
2138. Fache, M., "Improved microwave VTO in a to 8 package", Eur.
      Microwave Conf., 7th (MICROWAVE '77), Conf. Proc., 1977,
      p. 215-18.
2139. Fosco, P. and Hundt, M., "Ultra-reliable, low cost plastic
      packaging concept for microwave mixers", 27th Electronic Com-
      ponents Conference, 1977, p. 154, EEA80-35767.
2140. Krowne, C. M., Marki, F. A. and Crescenzi, E. J., Jr., "Ku-
      band receiver front-end using MIC technology", Dig. Tech. Pap.
      IEEE MTTS Int. Microwave Symp., 1977, p. 198-201.

2141. Lassan, C., "Microwave bonded packages", Electron, no. 119, 57-60 (1977), EEA81-10045.

2142. Lee, K. J., Parris, W. J. and Hosto, W. R., "Microwave integrated circuit techniques for multi-kilowatt amplifiers", Proceedings of the 1977 International Microelectronics Symposium, p. 185-8.

2143. Malmin, P. C., "Development and production of hybrid circuits for microwave radio links", Electrocompon. Sci. & Technol., 4 (2), 79-83 (1977), EEA81-5630.

2144. Meinel, H., Rembold, B. and Wiesbeck, W., "Optimisation of thin and thick film technology for hybrid microwave circuits", Electrocompon. Sci. & Technol., 4 (3-4), 143-6 (1977), EEA81-14335.

2145. Miley, J. E., "Impact-extruded MIC packaging", Electron. Packag. & Prod., 16 (3), 43-4, 46, 48 (1976).

2146. Nanbu, S., "MIC doppler module with output radiation normal to the substrate plane", IEEE Trans. Microwave Theory Tech., MTT-26 (1), 3-5 (1978).

2147. Nanbu, S., "New MIC doppler module", IEEE Trans. Microwave Theory Tech., MTT-26 (3), 192-6 (1978).

2148. Oxley, T. H., "Microwave integrated circuit techniques", GEC J. Sci. & Technol., 43 (1), 21-31 (1976), EEA80-3305.

2149. Oxley, T. H., "MICs using hybrid techniques for application at mm wavelengths", Microwave J., 19 (12), 46-7 (1976), EEA80-28630.

2150. Rehnmark, S., "Wide-band balanced line microwave hybrids", IEEE Trans. Microwave Theory Tech., MTT-25 (10), 825-30 (1977), EEA81-5448.

2151. Reindel, J., "Low dielectric substrates for MICs at millimeter wave frequencies", Proceedings of the 13th Electrical/Electronics Insulation Conference, 1977, p. 45-8, EEA81-14348.

2152. Rogers, C. B., Heeks, J. S., Barker, G. K. and Clarke, I. M., "A versatile solid-state MIC source module for Ku-band", Radio & Electron. Eng., 48 (1-2), 33-7 (1978), EEA81-19068.

2153. Rubin, D. and Saul, D., "MM-wave MICs use low value dielectric substrates", Microwave J., 19 (11), 35-9 (1976), EEA80-17044.

2154. Sherman, R. and McDowell, A. N., "Contemporary microwave packaging techniques", Microwave Syst. News, 5 (5), 75-81 (1975), EEA79-16191.

2155. Steddom, C., "Using thick-film techniques for microwave attenuators", Electron. Packag. & Prod., 18 (9), 61-2 (1978).

2156. Welke, R., "Qualification of the MIC technology for the OTS program", NTG-Fachber., no. 60, 196-200 (1977), German, EEA81-23643.

5.  Multilayer Ceramic

2157. Afanaseva, M. A., Kandyba, P. E. and Shutova, R. F., "Thick-film multilayer passive microcircuits", Mikroelektronika (Izd. "Sov. Radio"), 8, 114-28 (1975), Russian, CA85-185625.

2158. Aichelmann, F. J., Jarvela, R. A. and More, F. A., "Three dimensional multilevel circuit support package", IBM Tech. Disclosure Bull., 20 (11), 4349-50 (1978).

2159. Arnold, A. J., Coughlin, C. P. and Hardin, P. W., "Joining a
panel to features printed on the surface of a substrate", IBM
Tech. Disclosure Bull., 21 (2), 564-5 (1978).

2160. Brunner, J., Gonzales, F., Jr. and Leung, G. B., "Forming
green ceramic sheets on thin plastic films", IBM Tech. Disclosure Bull., 19 (8), 3022-3 (1977), EEA80-35276.

2161. Chirino, O. I., Hromek, J., Joshi, K. C., et al., "Multilayer
ceramic substrate structure", Patent USA 3999004, Publ.
December 1976.

2162. Cole, S. S. and Wellfair, G., "Achieving electrical stability
in hybrid multilayer dielectrics", Electrochemical Society
Fall Meeting, Extended Abstracts, 1975, p. 366, EEA79-16870.

2163. Crawford, D., Druschel, W. O., Martin, B. D. and Richards, L.,
"High density multilayer ceramic module", IBM Tech. Disclosure
Bull., 20 (11), 4771-3 (1978).

2164. Cuchelo, J. R. and Melvin, G. A., "Technique to minimize
device pad bulge during sintering a MLC module", IBM Tech.
Disclosure Bull., 21 (1), 134-5 (1978).

2165. Desai, K. S., "Curing multilayer ceramic green sheets", IBM
Tech. Disclosure Bull., 20 (9), 3423 (1978).

2166. Desai, K. S., "Technique for controlled paste filled via surface for preventing paste transfer at lamination for MLC
greensheets", IBM Tech. Disclosure Bull., 20 (9), 3438 (1978),
EEA82-606.

2167. Dougherty, W. E. and Greer, S. E., "Registration inspection
technique for multilayer structures", IBM Tech. Disclosure
Bull., 20 (11), 4353-4 (1978).

2168. Druschel, W. O. and Metreaud, C. G., "Integrated stacking
spacer for enhanced semiconductor packaging structure", IBM
Tech. Disclosure Bull., 21 (6), 2322-4 (1978).

2169. Dubetsky, D. J. and Meister, W. A., "Techniques for obtaining
uniform laminate density of unsintered multilayer ceramic
substrates", IBM Tech. Disclosure Bull., 20 (9), 3428 (1978).

2170. Dumaine, G., Durand, J., Lemoine, J. M. and Pottier, B.,
"Deposit of a preform for encapsulating multilayer ceramic
modules", IBM Tech. Disclosure Bull., 20 (4), 1470 (1977),
EEA81-23647.

2171. Dumaine, G., Lemoine, J. M. and Pottier, B., "Circuit package", IBM Tech. Disclosure Bull., 21 (7), 2852 (1978).

2172. Dzwilefsky, J. B., Kelly, L. F., Poulos, L. J., Rogers, C. T.
and Scalia, L., "Doctor/blade casting of ceramic greensheet",
IBM Tech. Disclosure Bull., 20 (8), 3083 (1978).

2173. Dzwilefsky, J. B., Kelly, L. F., Poulos, L. J. and Scalia,
L., "Monitoring technique and apparatus for greensheet casting", IBM Tech. Disclosure Bull, 20 (9), 3458-9 (1978).

2174. Ferrante, J. A., Melvin, G. E. and Urfer, E. N., "High conductivity lines in multilayer ceramic substrates", IBM Tech. Disclosure Bull., 21 (5), 1860-1 (1978).

2175. Galicki, A. and Sarkary, H., "Fabricating via openings in glass", IBM Tech. Disclosure Bull., 19 (12), 4580 (1977), EEA81-10242.

2176. Haddad, M. M., "Selectively electroplating pad terminals on an MLC substrate", IBM Tech. Disclosure Bull., 20 (9), 3443-4 (1978).

2177. Hardin, P. W., "Integral edge connector", IBM Tech. Disclosure Bull., 20 (11), 4346-8 (1978).

2178. Hayes, L. E. and Mayfield, R. E., "Improved multilayer performance with polyimide/glass laminates", Electron. Packag. & Prod., 17 (9), Pt. 1, 88-92 (1977).

2179. Hetherington, R. J., Milkovich, S. A. and Urfer, E. N., "Stand-offs for double sided screening of ceramic green sheets", IBM Tech. Disclosure Bull., 21 (5), 1862-3 (1978).

2180. Humenik, J. N., Kumar, A. H., Niklewski, J. B. and Tummala, R. R., "Multilayer ceramic substrates containing mullite", IBM Tech. Disclosure Bull., 20 (11), 4787 (1978).

2181. IBM Corp., "Ceramic green sheets", Patent UK 1475966, Publ. June 1977.

2182. Kaiser, H. D., "Multilayer module having optical channels therein", Patent USA 4070516, Publ. January 1978.

2183. Kim, Y. I., "Pin attachment structure for multilayer ceramic substrates", IBM Tech. Disclosure Bull., 20 (11), 4333 (1978).

2184. Kohler, H. and Meier, D., "On some problems of metal deposition without the use of external current supply in the case of ceramic multilayer enclosures (for monolithic ICs)", Hermsdorfer Tech. Mitt., 17 (47), 1496-9 (1977), German, EEA80-27980.

2185. Kutch, G., "Inspection apparatus for apertured green sheets", IBM Tech. Disclosure Bull., 20 (7), 2678 (1977).

2186. Lehman, H. S. and Tummala, R. R., "Eliminating via-to-via cracks in multilayer alumina ceramic substrates", IBM Tech. Disclosure Bull., 20 (1), 141 (1977), EEA81-13595.

2187. Loehndorf, D., "Electroplating process for forming heavy gold plating on multilayer ceramic substrates", IBM Tech. Disclosure Bull., 20 (5), 1740 (1977).

2188. Lussow, R. O., "Internal capacitors and resistors for multilayer ceramic modules", IBM Tech. Disclosure Bull., 20 (9), 3436-7 (1978), EEA82-605.

2189. Magdo, S., "Low inductance module", IBM Tech. Disclosure Bull., 21 (5), 1895-7 (1978).

2190. Master, R. N., McMillan, P. W. and Tummala, R. R., "Use of precrystallized glass powder for production of multilayer circuit boards in glass ceramic", IBM Tech. Disclosure Bull., 21 (3), 1008-13 (1978).

2191. Melvin, G. E., Milkovich, S. A. and Urfer, E. N., "Via place-
      ment in multilayer ceramic substrates", IBM Tech. Disclosure
      Bull., 21 (3), 1021-2 (1978).
2192. Morton, R. M., North, W. D. and Rymaszewski, E. J., "High-
      performance AC chip contactor", IBM Tech. Disclosure Bull.,
      18 (3), 749-50 (1975), EEA79-1002.
2193. Sarkary, H. G. and Tewari, V., "Via holes in quartz coatings
      [LSI]", IBM Tech. Disclosure Bull., 19 (11), 4152-3 (1977),
      EEA81-10368.
2194. Schmeckenbecher, A. F., "Metalizing ceramic green sheets",
      Patent USA 3948706, Publ. April 1976.
2195. Swiss, W. R. and Young, W. S., "Accelerated sintering for a
      green ceramic sheet", Patent USA 4039338, Publ. August 1977.
2196. Wade, J. F., "Advances in ceramic chip carriers and multi-
      layer substrate technologies", Reliability Physics 16th Annual
      Proceedings, 1978, p. 130-1, EEA81-48757.
2197. Wolfe, G., "Technique produces ceramic substrates with
      0.002'/' camber", Circuits Manuf., 15 (10), 38, 40, 42 (1975),
      EEA79-12983.

### 6.  Optoelectronics

2198. Archey, W. B., "Optical waveguides built into modules", IBM
      Tech. Disclosure Bull., 20 (2), 537-8 (1977).
2199. Biard, J. R., "Optoelectronic devices packaged for fiber
      optics application", Report AD-A025905/1SL, Spectronics, Inc.,
      Richardson, Tex., 1976, 131 pp.
2200. Crow, J. D., Comerford, L. D., Laff, R. A., Brady, M. J. and
      Harper, J. S., "GaAs laser array source package", Opt. Lett.,
      1 (1), 40-2 (1977), EEA80-41944.
2201. Fang, F. F. and Hung, R. Y., "Merging semiconductor optoelec-
      tronics with silicon technology", IBM Tech. Disclosure Bull.,
      19 (10), 3959-60 (1977), EEA81-5633.
2202. Gretch, G. E. and Meyer, R. A., "Visual aid for signaling deaf
      students", Proceedings of the 1977 International Microelec-
      tronics Symposium, p. 297-9.
2203. Hamilton, M. C., Wille, D. A. and Miceli, W. J., "Integrated
      optical rf spectrum analyzer", Opt. Eng., 16 (5), 475-8 (1977).
2204. Kalita, W., Kusy, A. and Paszczynski, S., "New possibilities
      of thick film technology in rutenate resistors, switches,
      transistors and optoelectronic devices", Rozpr. Electrotech.,
      22 (4), 831-54 (1976), Polish, EEA80-9199.
2205. Maeda, M., Nagano, K., Tanaka, M. and Chiba, K., "Buried-
      heterostructure laser packaging for wideband optical trans-
      mission systems", IEEE Trans. Commun., COM-26 (7), 1076-81
      (1978), EEA81-43511.
2206. Marshall, J. and Rode, F., "Solder bump interconnected,
      multiple chip, thick film hybrid for a 40-character alpha-
      numeric LCD application", Solid State Technol., 22 (1),
      87-93 (1979).

2207. Piggin, B. P. and Proudman, A., "Absorbent package for elec-
      trochromic display", IBM Tech. Disclosure Bull., 21 (5),
      2046 (1978).
2208. Price, S. J., Johnson, R. L. and Chapman, J. F., "Encapsula-
      tion of optoelectronic devices for severe environments", 26th
      Electronic Components Conference, 1976, p. 386-9, EEA79-38021.
2209. Price, S. J., Johnson, R. L. and Chapman, J. F., "For reli-
      able, severe environment performance, encapsulate LED's with
      a clear silicone", Insul./Circuits, 23 (11), 53-5 (1977).
2210. Sigmund, M., "Integrated optical assemblies produced by
      plastic mouldings, for use in the electronic industry",
      Elektroniker, 15 (2), EL23-6 (1976), German, EEA79-19576.
2211. Slayton, I. B., Estapa, D. J., Jones, J. R. and Patisaul, C.
      R., "Fiber optics design aid package", Report AD-A040772/6SL,
      Harris Corp., Melbourne, Fla., 1977, 201 pp.
2212. Smith, M. R., Bell, I. J. and Krystich, M., "Electro-optical
      thin films for data storage and display", SPIE Semin. Proc.
      Vol. 123, Opt. Storage Mater. and Methods, 1977, p. 113-12.
2213. Speer, R. S., "Optoelectronic devices packaged for fiber
      optics applications", Report AD-A014652/2SL, Spectronics,
      Inc., Richardson, Tex., 1975, 42 pp.
2214. Trasatti, A. J., "Packaging an electronic display unit for
      the B-1 aircraft", Electron. Packag. & Prod., 16 (12), 106-8
      (1976).
2215. Vance, R. D. and Griffith, W. F., "Packaging CRT displays in
      near-field environments", Electron. Packag. & Prod., 18 (10),
      125-6, 128, 130, 132 (1978).

### 7. Polyimide

2216. Abolafia, O. R. and Rasile, J., "Use of polyimide to obtain
      a smooth surface", IBM Tech. Disclosure Bull., 20 (8), 3020
      (1978).
2217. Anon., "Polyimide flexible circuits reduced wiring costs by
      50%", Des. Eng., 25 (1978), EEA81-32989.
2218. Berger, A., "Polyimide containing silicones as protective
      coating on a semiconductor device", Patent USA 4030948,
      Publ. June 1977, CA87-61588.
2219. De Brebisson, M. and Tessier, M., "Preliminary studies on the
      use of polyimides in semiconductor technology", Vide, 76
      (183), 195-7 (1976), French.
2220. Ellozy, A. R., Knapp, E. C. and Roush, W. B., "Coating and
      process to minimize water adsorption on polyimide laminate",
      IBM Tech. Disclosure Bull., 20 (10), 3882 (1978).
2221. Fusaro, R. L., "Effect of atmosphere and temperature on wear,
      friction, and transfer of polyimide films", ASLE Prepr. 1976,
      ASME/ASLE Lubr. Conf., Pap. 76-LC-1B-1, 8 pp.
2222. Hayes, L. E. and Mayfield, R. E., "Polyimide - an emerging
      printed wiring laminate material", IEEE Trans. Electr. Insul.,
      EI-13 (2), 105-9 (1978), EEA81-23583.

2223. Hayes, L. E. and Mayfield, R. E., "Multilayer boards of poly-
      imide/glass provide high reliability in rigorous environments",
      Insul./Circuits, 24 (4), 35-8 (1978), EEA81-42114.
2224. Hitchner, J. E. and O'Rourke, G. D., "Polyimide layers having
      tapered via holes", IBM Tech. Disclosure Bull., 20 (4), 1384
      (1977), EEA81-23898.
2225. Kouchinsky, J., "Using dynamic dielectric analysis as an in-
      strumental technique, in defining polymide composite bonding
      problems", Natl. SAMPE Symp. and Exhib., 22nd, 1977, p. 618-24.
2226. Licari, J. J., Varga, J. E. and Bailey, W. A., "Polyimide
      supported micro-ramps for high density circuit interconnec-
      tion", Solid State Technol., 19 (7), 41-4 (1976), EEA79-37622.
2227. Miller, S. C., "Passivating thin film hybrids with polyimide",
      Circuits Manuf., 17 (4), 54, 56, 58 (1977), EEA81-5627.
2228. Mukai, K., Saiki, A., Yamanaka, K., Harada, S. and Shoji, S.,
      "Planar multilevel interconnection technology employing a
      polyimide", IEEE J. Solid-State Circuits, SC-13 (4), 462-7
      (1978), CA89-172482.
2229. Progar, D. J. and St. Clair, T., "Preliminary evaluation of
      a novel polyimide adhesive for bonding titanium and reinforced
      composites", National SAMPE Technical Conference, Vol. 7,
      1975, p. 53-67.
2230. Rapoport, N. R., "Removing dual insulating layers over a me-
      tallic layer", IBM Tech. Disclosure Bull., 20 (8), 3004-5
      (1978).
2231. Sacher, E. and Woods, J. J., "Improving bonding strength of
      polyimide to silicon wafers", IBM Tech. Disclosure Bull., 20
      (6), 2161 (1977).
2232. Schneier, R., Braswell, T. V. and Vaughn, R. W., "Feasibility
      demonstration for electroplating ultra-thin polyimide film",
      Report N78-33266/5SL, TRW Defense and Space Systems Group,
      Redondo Beach, CA, 1978, 25 pp.
2233. Sroog, C. E., "Polyimides", J. Polym. Sci. Macromol. Rev.,
      11, 161-208 (1976).

## 8.  Others

2234. Anon., "Burroughs' E-4B CPU packaged for severe environment",
      Electron. Packag. & Prod., 17 (9), Pt. 1, 81-2, 84 (1977).
2235. Bachmann, P. and Dujmovic, B., "Sitor thyristor assemblies
      with disc-type thyristors", Siemens Rev., 44 (10), 462-5 (1977).
2236. Bauer, J. A., "Using chip carriers for high density packaging",
      Electron. Packag. & Prod., 17 (10), 85-6, 88, 91 (1977).
2237. Brown, V. L., "LAMPAC: an alterative to MLBs", Circuits
      Manuf., 18 (10), 22, 24, 26, 28 (1978).
2238. Buchoff, L. S., Dalamangas, C. A. and Piccirillo, T. P., "Con-
      ductive rubber connectors for leadless electronic packages",
      International Microelectronics Conference, Proceedings of the
      Technical Program, 1975, p. 48-53.

2239. Cain, O. J. and Ordonez, J., "Circuit module with removable cap", IBM Tech. Disclosure Bull., 19 (5), 1803 (1976).

2240. Coe, J. E. and Oldham, W. G., "Enter the 16384-bit RAM", Electronics, 49 (4), 116-22 (1976), CCA11-15060.

2241. Crawford, D. J., "High power electronic package", IBM Tech. Disclosure Bull., 20 (11), 4393-8 (1978).

2242. Doo, V. Y. and Sachar, K. S., "High performance package for memory", IBM Tech. Disclosure Bull., 21 (2), 585-6 (1978).

2243. Ferraro, A. J., Larnerd, J. D., Tompkins, T. E., Tomsa, D. R. and Zucconi, T. D., "Hybrid cloth for making low alpha laminates", IBM Tech. Disclosure Bull., 20 (9), 3396-7 (1978), EEA82-651.

2244. Forster, J. H., "Polychip DIPs: quick-turnaround custom circuit designs", Bell Lab. Rec., 55 (8), 209-14 (1977).

2245. Hatfield, W. and Wicher, D., "Metal board technologies", Circuits Manuf., 18 (7), 7-8, 10, 12 (1978).

2246. Hatfield, W. B. and Wicher, D. P., "A microelectronics packaging technology for consumer product applications", Proceedings of the 28th Electronics Components Conference, 1978, p. 271-6, EEA81-37929.

2247. Hilson, D. G. and Johnson, G. W., "New materials for low cost thick film circuits", Solid State Technol., 20 (10), 49-54, 75 (1977), EEA81-19161.

2248. Huss, R. H., "Navy standard electronic modules program approach to microprocessor/microcomputer standardization", 14th IEEE Computer Society International Conference, COMPCON 77, Digest of Papers, p. 267-70.

2249. Katyl, R. H., "Monolithic ultrasound array", IBM Tech. Disclosure Bull., 20 (6), 2183-4 (1978).

2250. Lyman, O., "Growing pin count is forcing LSI package changes", Electronics, 50 (6), 81-91 (1977), EEA80-31390.

2251. Markstein, H. W., "Semiconductor devices and the question of pluggability", Electron. Packag. & Prod., 17 (10), 61-2, 64, 66 (1977).

2252. Markstein, H. W., "Packaging for military and aerospace applications", Electron. Packag. & Prod., 18 (2), 50-5 (1978), EEA82-102.

2253. Mauriello, A. J., "Packaging of articles", Tech. Dig., no. 46, 27-8 (1977), EEA81-13298.

2254. May, D. S. and Piller, D. W., "Integrated power-logic board", IBM Tech. Disclosure Bull., 20 (6), 2372-3 (1977), EEA82-109.

2255. Miksic, B. A., "Environmental protection using vapour phase inhibitors", Electron. Prod. Methods & Equip., 4 (10), 22, 24 (1975-76), EEA79-19574.

2256. Moore, D. W., "Packaged thyristor circuits gain ground", Electr. Times, no. 4364, 7, 11 (1976), EEA79-20464.

2257. Pollack, H. W., "Flexible circuits a packaging product whose time has come", 8th Annual Connector Symposium, 1975, p. 265-74, EEA79-3735.

2258. Russell, D., "10k e.c.l. — the systematic choice", New Elec-
      tron., $\underline{8}$ (24), 37-8, 42 (1975), CCA11-9582.
2259. Schweihs, J. J., "Electronic circuit card and board packag-
      ing", IBM Tech. Disclosure Bull., $\underline{19}$ (6), 2204-6 (1976),
      EEA80-28617.
2260. Settle, R. E., Jr., "A new family of microelectronics packages
      for avionics", Solid State Technol., $\underline{21}$ (6), 54-8 (1978),
      EEA81-45187.
2261. Staley, W. W., "Modular packaging approaches", Report AD-
      A048205/9SL, Westinghouse Electric Corp., Baltimore, Md.,
      1977, 124 pp.
2262. Taylor, C. H., "The military use of plastic encapsulated semi-
      conductor devices", Proceedings of the Symposium on Plastic
      Encapsulated Semiconductor Devices, 1976, p. 1/1-13, EEA80-6603.
2263. Tiefert, K. H., "Chip configuration of high temperature pres-
      sure contact packaging of Schottky barrier diodes", Patent
      USA 3961350, Publ. June 1976, CA85-71349.
2264. Whitelaw, D., "Packaged resistors", Aust. Electron. Eng., $\underline{9}$
      (2), 33 (1976), EEA79-32663.
2265. Whitelaw, D., "Thin film — new applications", Microelectron.
      & Reliab., $\underline{16}$ (4), 319-21 (1977), EEA81-10039.

XXII.   RELIABILITY
        1.  General

2266. Alexanian, I. T. and Brodie, D. E., "Method for estimating
      the reliability of ICs", IEEE Trans. Reliab., R-$\underline{26}$ (5),
      359-61 (1977).
2267. Anderson, L. K., "Some package reliability implications of
      current trends in large scale silicon integrated circuits",
      Reliability Physics Symposium, 16th Annual Proceedings,
      1978, p. 121-3, EEA81-48754.
2268. Anon., "Uncovering hybrid failures", Circuits Manuf., $\underline{16}$ (6),
      62-4 (1976), EEA79-37708.
2269. Anon., "What makes semicons fails? (failure analysis tip)",
      Circuits Manuf., $\underline{16}$ (6), 66, 68 (1976), EEA79-37523.
2270. Anon., "Expanding SEM usefulness by avoiding harmful effects",
      Circuits Manuf., $\underline{17}$ (12), 22, 24 (1977).
2271. Anon., "How to improve microcircuit reliability. II. Detect-
      ing microcircuit packages. PIN test helps spot contaminants
      nondestructively", Circuits Manuf., $\underline{17}$ (6), 28-32 (1977),
      EEA81-14332.
2272. Avery, L. R., "Improvements in the reliability of integrated
      circuits", Proceedings of the Symposium on Plastic Encapsulated
      Semiconductor Devices, 1976, p. 13/1-17, EEA80-6691.
2273. Bart, J. J., "Scanning electron microscopy for complex micro-
      circuit analysis", 16th Annual Proceedings Reliability Physics
      Symposium, 1978, p. 108-11.

2274. Beall, J. R. and Hamiter, L., Jr., "EBIC – a valuable tool for semiconductor evaluation and failure analysis", 15th Annual Proceedings Reliability Physics Symposium, 1977, p. 61-9.

2275. Bobbio, A. and Saracco, O., "A modified reliability expression for the electromigration time-to-failure", Microelectron. & Reliab., 14 (5-6), 431-3 (1975), EEA79-24008.

2276. Ebel, G. H., "Failure analysis techniques applied in resolving hybrid microcircuit reliability problems", 15th Annual Proceedings Reliability Physics Symposium, 1977, p. 70-81, EEA80-41114.

2277. Edfors, H. C., "Increasing integrated circuit product reliability through failure analysis: The role of the independent laboratory", IEEE Trans. Manuf. Technol., MFT-6 (1), 7-9 (1977), EEA80-28900.

2278. Fischer, F. and Fellinger, J., "Failure analysis of thin-film conductors stressed with high current density by application of Matthiessen's rule", 15th Annual Proceedings Reliability Physics Symposium, 1977, p. 250-6.

2279. Gallace, L. J. and Pujol, H. L., "Failure mechanisms in COS/MOS integrated circuits", Electron. Eng., 48 (586), 65, 67, 69 (1976).

2280. George, A. H., "The reliability, testing and evaluation of hybrid microcircuits", Electrocompon. Sci. & Technol., 4 (3-4), 213-17 (1977), EEA81-14338.

2281. Gonzales, A. J. and Powell, M. W., "Improved integrated circuit failure analysis using SEM video processing", J. Vac. Sci. & Technol., 15 (3), 837-40 (1978), EEA81-42440.

2282. Hawkins, R., "Reliability improvements in commercial/industrial grade integrated circuit devices", IEEE Trans. Manuf. Technol., MFT-5 (3), 58-61 (1976).

2283. Himmel, R. P., "How to improve microcircuit reliability. I. Why hybrids fail. Failure analysis of low- and high-power microcircuits", Circuits Manuf., 17 (6), 22, 24, 26, 28-9 (1977), EEA81-14331.

2284. Kovac, M. G., Chleck, D. and Goodman, P., "New moisture sensor for "in-situ" monitoring of sealed packages", 15th Annual Proceedings Reliability Physics Symposium, 1977, p. 85-91.

2285. Krivosheykin, A. V., "Optimal synthesis of the tolerances on the elements of thin-film hybrid circuits", Radio Eng. Electron. Phys., 21 (11), 93-9 (1976).

2286. Kubo, Y., Ohata, K., Matsumura, K. and Suzuki, H., "Reliability management for Hitachi integrated circuits", Hitachi Rev., 24 (6), 268 (1975), EEA79-1000.

2287. Kusko, A., Knutrud, T. and Cain, J. J., "Designing reliability into power circuits - 2.", Electronics, 49 (6), 101-5 (1976).

2288. Lee, S. M., "Advanced instrumental-failure analytical techniques", Electron. Packag. & Prod., 17 (10), 72-4, 80-3 (1977).

2289. Livesay, B. R., "The reliability of electronic devices in storage environments", Solid State Technol., 21 (10), 63-8 (1978).

2290. Lynch, J. T., Hatfield, J. A. and Brydon, G. M., "The role of customer QA in reducing hybrid reliability problems", Microelectron. & Reliab., 16 (4), 523-6 (1977), EEA81-10054.

2291. Nakagawa, T., "Reliability analysis of standby repairable systems when an emergency occurs", Microelectron. & Reliab., 17 (4), 461-4 (1978), EEA81-41432.

2292. Parkinson, D. B. and Martin, P., "The effect of quality control procedures on component reliability", IEEE Trans. Reliab., R-25 (5), 349-51 (1976), EEA80-12546.

2293. Peck, D. S., "New concerns about integrated circuit reliability", IEEE Trans. Electron Devices, ED-26 (1), 38-42 (1979).

2294. Petrikovits, L. and Goblos, I., "Reliability testing of thick film circuit models", Hiradastechnika, 28 (4), 101-9 (1977), Hungarian, EEA81-5623.

2295. Platteter, D., "Basic integrated circuit failure analysis techniques", 14th Annual Proceedings Reliability Physics Symposium, 1976, p. 248-55, EEA80-17260.

2296. Poyer, T. H. and Schoonmaker, T. D., "MLA reliability assurance - a continuing program", 1978 Proceedings Annual Reliability and Maintainability Symposium, p. 428-33, EEA81-33005.

2297. Schiller, N., "Reliability of monolithic integrated circuits", Radio Fernsehen Elektron., 25 (5), 150-2 (1976), German, EEA79-28687.

2298. Schnable, G. L., "State of the art in semiconductor materials and processing for microcircuit reliability", Solid State Technol., 21 (10), 69-73 (1978).

2299. Schroen, W. H., "Process testing for reliability control", 16th Annual Proceedings Reliability Physics Symposium, 1978, p. 81-7.

2300. Sletten, E. and Olesen, S. T., "Evaluation of the technology of semiconductor devices - 2. Procedure for methodical failure analysis and description of typical failures in encapsulation, bonding, metallization, passivation layer and bulk silicon", Elektronikcentralen Rep. ECR, no. 51, 76 pp. (1975).

2301. Strecker, A., "Quality assurance applied to packaging", Qual. & Zuverlassigkeit, 20 (10), 218-22 (1975), German, EEA79-3731.

2302. Thielmann, J. N., "Reliability considerations in the use of integrated circuit packaging systems in an automotive environment", SAE Prepr. No. 770229, 1977, 34 pp.

2303. Turner, T. E., "Microcircuit device reliability, hybrid circuit data", Report AD-A040303/0SL, Reliability Analysis Center, Griffiss AFB, N. Y., 1977, 409 pp.

2304. Valko, I. P., "Unsolved problems in the reliability theory of integrated circuits", Acta Tech. Acad. Sci. Hung., 80 (1-2), 41-9 (1975), EEA79-1033.

2305. Van Keuren, E., Hendrickson, R. and Magyarics, R., "Circuit failure thresholds due to transient induced stresses", 1st Symposium and Technical Exhibition on Electromagnetic Compatibility, Proceedings, 1975, p. 500-5.

2306. Weglein, R. D., Wilson, R. G. and Bonnell, D. M., "Scanning
      acoustic microscopy - application to fault detection", 15th
      Annual Proceedings Reliability Physics Symposium, 1977,
      p. 37-43.
2307. Weirick, L. J., "Prevention of liquid-metal embrittlement and
      stress-corrosion cracking in Kovar leads", Solid State Tech-
      nol., 19 (6), 55-61 (1976).

## 2. Bond Failure

2308. Andrade, A. D., "Preventing solder joint failure in planar-
      mounted hybrid ICs", Insul./Circuits, 22 (4), 27-32 (1976),
      EEA79-28706.
2309. Bartsch, P. and Stojanoff, H., "Prediction of soldering errors
      in electronic assemblies", Fernmeldetechnik, 16 (6), 193-7
      (1976), German, EEA80-19712.
2310. Basseches, H. and D'Altroy, F. A., "Shear mode wire failures
      in plastic-encapsulated transistors", IEEE Trans. Components,
      Hybrids & Manuf. Technol., CHMT-1 (2), 143-7 (1978).
2311. Baun, W. L., "Scanning electron microscopy, ion scattering
      and secondary ion mass spectrometry to characterize apparent
      'adhesive' failure in an adhesive bond", Report AD-A025615/
      6SL, Air Force Materials Lab., Wright-Patterson AFB, Ohio,
      1976, 44 pp.
2312. Baun, W. L., "Experimental methods to determine locus of fail-
      ure and bond failure mechanism in adhesive joints and coating-
      substrate combinations", ASTM Spec. Tech. Publ. No. 640;
      Adhes. Meas. of Thin Films, Thick Films, and Bulk Coat., 1976,
      p. 41-53, Publ. 1978.
2313. Bulwith, R. A., "Failure analysis of solder joints", Insul./
      Circuits, 24 (2), 19-23 (1978).
2314. Bushmire, D. W. and Holloway, P. H., "Correlation between
      bond reliability and solid phase bonding techniques for con-
      taminated bonding surfaces", Report SAND-75-5600, Sandia
      Labs., Albuquerque, N. Mex., 1975, 6 pp.
2315. Caruso, S. V., Kinser, D. L., Graff, S. M. and Allen, R. V.,
      "The relationship between reliability and bonding techniques
      in hybrid microcircuits", Report N75-32322/0SL, NASA, Mar-
      shall Space Flight Center, Huntsville, Ala., 1975, 28 pp.
2316. Cohn, E., "Effects of wire quality and capillary maintenance
      on bonding reliability (semiconductor device manufacture)",
      Solid State Technol., 18 (9), 31-6 (1975), EEA79-1070.
2317. Dais, J. L. and Howland, F. L., "Beam fatigue as a failure
      mechanism of gold beam lead and TAB encapsulated devices",
      Proceedings of the 1977 International Microelectronics Sym-
      posium, p. 100-5.
2318. Dietz, R. L., "New gold alloy for aluminium bonding in MOS
      hermetic packages", 27th Electronic Components Conference,
      1977, p. 36-73, EEA80-35969.

2319. Dreyer, G. A., Koudounaris, A. and Pratt, I. H., "The reliability of soldered or epoxy bonded chip capacitor interconnections in hybrids", IEEE Trans. Parts, Hybrids & Packag., PHP-13 (3), 218-24 (1977), EEA81-5634.

2320. Dunn, B. D., "Producing highly reliable joints for spacecraft electronics. I.", Electron. Prod., 7 (3), 27, 29-30 (1978), EEA81-45168.

2321. Ewell, G. J., "Reliability problems in reflow soldering of Ag- and Pd/Ag-terminated chip components", 27th Electronic Components Conference, 1977, p. 206-11, EEA80-35967.

2322. Ferraro, A. J., Larnerd, J. D. and Zucconi, T. D., "Joint reliability using a plated copper stud", IBM Tech. Disclosure Bull., 20 (5), 1722 (1977), EEA81-37931.

2323. Guidici, D. C., "Ribbon wire versus round wire reliability for hybrid microcircuits", Electron. Inf. & Plann., 3 (8), 664-8 (1976), EEA79-46547.

2324. Gonzalez, H. M., "Component mounting methods and their effect on solder joint cracking", Insul./Circuits, 23 (11), 19-24 (1977), EEA81-13277.

2325. Gray, H. F., "Surface segregation of non-bonding impurities in gold-silicon preforms", 15th Annual Proceedings Reliability Physics Symposium, 1977, p. 272-5, EEA80-41121.

2326. Hakim, E. B., "Reliability considerations in the application of plastic semiconductors", 12th Electrical/Electronics Insulation Conference, 1975, p. 68, EEA79-37375.

2327. Hartley, R. L., "Electrical shorts result of using backwards bonding techniques", Electron. Packag. & Prod., 18 (1), 213-14 (1978).

2328. Hinderer, A., Lessmann, B. and Mueller, H., "Method of reliably opening fusible links", IBM Tech. Disclosure Bull., 20 (4), 1494 (1977), EEA81-24038.

2329. Hoffman, M. and Wrinn, J., "Diagnosing faults on LSI bus lines", Circuits Manuf., 18 (12), 14, 18 (1978).

2330. Horowitz, S. J., Gerry, D. J. and Cote, R. E., "Connecting to gold thick film conductors: performance and failure mechanisms", Proc. of the Tech. Program, Int. Microelectron. Conf., 1977, p. 47-56.

2331. Hubregtse, J., "Gold alumina and gold quartz bonds for microwave integrated circuits", ATR Aust. Telecommun. Res., 12 (1), 35-42 (1978).

2332. James, K., "Reliability study of wire bonds to silver plated surfaces", IEEE Trans. Parts, Hybrids & Packag., PHP-13 (4), 419-25 (1977), EEA81-10062.

2333. Jellison, J. L., "Susceptibility of microwelds in hybrid microcircuits to corrosion degradation", 13th Annual Proceedings of Reliability Physics Symposium, 1976, p. 70-9, EEA79-46575.

2334. Kanamori, S. and Takahashi, M., "Silicon die bond failure induced by nickel-silicon intermetallic formation", Rev. Electr. Commun. Lab., 23 (5-6), 552-8 (1975).

2335. Kang, S. K., Zommer, N. D., Feucht, D. L. and Heckel, R. W., "Thermal fatigue failure of soft-soldered contacts to silicon power transistors", IEEE Trans. Parts, Hybrids & Packag., PHP-13 (3), 318-21 (1977).

2336. Kashar, L., "Failure analysis of EB welded metal hybrid packages", Advanced Techniques in Failure Analysis Symposium, 1976, p. 80-5, EEA79-46551.

2337. Khadpe, S., "Corrosion characteristics of passivated and un-passivated Al-to-PtAg wire bonds under accelerated temperature-humidity conditions", Proceedings of the 13th Electrical/Electronics Insulation Conference, 1977, p. 61-4, EEA81-14361.

2338. Kinser, D. L., "Discrete component bonding and thick film materials study", Report N77-14483/0SL, Vanderbilt Univ., Nashville, Tenn., 1976, 81 pp.

2339. Kinser, D. L., Vaughan, J. G. and Graff, S. M., "Reliability of soldered joints in thermal cycling environments", Electron. Packag. & Prod., 16 (5), 61-2, 64, 66, 68 (1975), EEA79-41580.

2340. Kobayashi, M., "Reliability of soldered connections in single ended circuit components", Weld. J., 54 (10), 363-9 (1975), CA84-98459.

2341. Kossowsky, R. and Robinson, A. I., "Investigation into fail-ures of Al wires bonded to Au metallization in microsubstrates", 14th Annual Proceedings Reliability Physics Symposium, 1976, p. 75-81, EEA80-17047.

2342. Macha, M. and Theil, R. A., "High reliability bond program using small diameter aluminum wire", Report N75-32462/4SL, General Dynamics/Electronics, San Diego, Calif., 1975, 55 pp.

2343. Munford, J. W., "Influence of several design and material variables on the propensity for solder joint cracking", IEEE Trans. Parts, Hybrids & Packag., PHP-11 (4), 296-304 (1975).

2344. Newsome, J. L., Oswald, R. G. and Rodrigues de Miranda, W. R., "Metallurgical aspects of aluminum wire bonds to gold metallization", 14th Annual Proceedings Reliability Physics Symposium, 1976, p. 63-74, EEA80-17046.

2345. Palmer, D. W. and Ganyard, F. P., "Aluminum wire to thick-film connections for high-temperature operation", IEEE Trans. Com-ponents, Hybrids & Manuf. Technol., CHMT-1 (3), 219-22 (1978), EEA81-49511.

2346. Peckinpaugh, C. J., "Adhesion loss in solderable thick film conductor systems as a result of thermal shock", Insul./Circuits, 23 (3), 35-8 (1977).

2347. Penczek, E. S., "Blue plague discoloration of reflowed solder before assembly", Circuits Manuf., 16 (12), 36, 38-43 (1976), EEA80-24599.

2348. Perkins, K. L. and Licari, J. J., "Investigation of low cost, high reliability sealing techniques for hybrid microcircuits", Report N76-27364/8SL, Rockwell International Corp., Anaheim, Calif., 1976, 67 pp.

2349. Pietrucha, B. M. and Reiss, E. M., "Reliability of epoxy as a die attach in digital and linear integrated circuits", RCA Eng., 21 (5), 68-71 (1976).

2350. Rodrigues de Miranda, W. R., "How storage at elevated temperatures affects aluminium wire to gold metallization bonds", Insul./Circuits, 22 (5), 25-8 (1976), EEA79-31906.

2351. Sinha, A. K., "Failure mechanism in wire bonds", J. Inst. Eng. Electron. & Telecommun. Eng. Div., 58 (pt. ET-1), 18-22 (1977), EEA81-23903.

2352. Thwaites, C. J., "In-built reliability of soldered joints in electronic assemblies", Electron, no. 124, 62, 67, 71, 74 (1977), EEA81-18344.

2353. Thomas, R. E., Winchell, V., James, K. and Scharr, T., "Plastic outgassing induced wire bond failure", 27th Electronic Components Conference, 1977, p. 182-7, EEA80-35966.

2354. Woodward, J. P., "Does appearance indicate solder joint reliability?", Circuits Manuf., 18 (1), 44-53 (1978), EEA81-45167.

3.   Circuitry Failure

2355. Breen, J. E., Toledo, E. and Sukis, D., "The annealing characteristics of one mil aluminum/1% silicon wire used in ICs", Electrochemical Society Fall Meeting, Extended Abstracts, 1975, p. 452-4, EEA79-16900.

2356. Cerofolini, G. F. and Rovere, C., "The role of water vapour in the corrosion of microelectronic circuits", Thin Solid Films, 47 (2), 83-94 (1977), EEA81-14308.

2357. Comizzoli, R. B., "Aluminium corrosion in the presence of phosphosilicate glass and moisture", RCA Rev., 37 (4), 483-90 (1976), EEA80-24975.

2358. Dozier, A. W., McCormac, D. E., Engguist, R. D. and Clarke, R., "Failure mechanisms in discrete semiconductor devices", Advanced Techniques in Failure Analysis Symposium, 1976, p. 35-47, EEA79-45923.

2359. Dreyer, G. A., Koudounaris, A. and Pratt, I. H., "The reliability of soldered or epoxy bonded chip capacitor interconnections on hybrids", IEEE Trans. Parts, Hybrids & Packag., PHP-13 (3), 218-24 (1977).

2360. Fresh, D. L., "Reliability of interconnections on microcircuits", Proceedings of the 1975 Annual Reliability and Maintainability Symposium, p, 568-72, EEA79-46395.

2361. Fuss, F. N., Hartwig, C. T. and Morabito, J. M., "Corrosion of solder-coated TiPdAu thin film conductors in a moist chlorine atmosphere", Thin Solid Films, 43 (1-2), 189-213 (1977), EEA80-35906.

2362. Griffin, D. D. and Waldman, D. P., "Integrated-circuit package structure for preventing sulfur corrosion", IBM Tech. Disclosure Bull., 19 (11), 4147 (1977), EEA81-10032.

2363. Haddad, M. M., Harvilchuck, J. M. and Schmeckenbecher, A. F., "Process for preventing chip pad corrosion", IBM Tech. Disclosure Bull., 19 (12), 4581 (1977), EEA81-10060.

2364. Hall, E. L., Koblinski, A. N. and Gragg, J. E., "Beryllium - a high reliability metallization", 14th Annual Proceedings Reliability Physics Symposium, 1976, p. 48-54, EEA80-17183.

2365. Herman, D. S., Kossowsky, R. and Johnson, E. W., "'Bubble' formations on aluminium multilevel metal structures on sapphire substrates", Electrochemical Society Spring Meeting, Extended Abstracts, 1975, p. 115-16, EEA79-1109.

2366. Lee, F. F. M., "Reliability of clad metal inlays for electrical connectors", IEEE Trans. Parts, Hybrids & Packag., PHP-13 (1), 61-7 (1977).

2367. Holmes, P. J., "Reliability and failure mechanisms [in thick film technology]", In: Handb. Thick Film Technol., Holmes, P. J. and Loasby, R. G., (Eds.), Electrochem. Publications, Ltd.: Ayr, Scot., p. 380-98 (1976), CA88-97822.

2368. Kinsborn, E., Melliar-Smith, C. M. and English, A. T., "Failure of small thin-film conductors due to high current-density pulses", IEEE Trans. Electron Devices, ED-26 (1), 22-6 (1979).

2369. Kossowsky, R., Pearson, R. C. and Christovich, L. T., "Corrosion of In-base solders", 16th Annual Proceedings Reliability Physics Symposium, 1978, p. 200-6, EEA81-48750.

2370. Lazarova, Z., Vaces, K. and Mateva, I., "On the stability of thin film circuits used in electronic calculators", Elektro Prom.-st. & Priborostr., no. 3, 109-10 (1977), Bulgarian, EEA81-14315.

2371. McAteer, O. J., "Pulse evaluation of integrated circuit metallization as an alternative to SEM", 15th Annual Proceedings Reliability Physics Symposium, 1977, p. 217-24, EEA80-41429.

2372. Miller, R. J., "Electromigration failure under pulse test conditions", 16th Annual Proceedings Reliability Physics Symposium, 1978, p. 241-7.

2373. Olberg, R. C. and Bozarth, J. L., "Factors contributing to the corrosion of the aluminum metal on semiconductor devices packaged in plastics", Microelectron. & Reliab., 15 (6), 601-11 (1976).

2374. Panousis, N. T., Wonsiewicz, B. C. and Condra, L. W., "Oxygen embrittlement of copper leads", IEEE Trans. Parts, Hybrids & Packag., PHP-13 (2), 127-32 (1977), EEA80-32248.

2375. Piacentini, G. F. and Minelli, G., "Reliability of thin film conductors and air gap crossovers for hybrid circuits: tests, results and design criteria", Microelectron. & Reliab., 15 (5), 451-8 (1976), EEA80-6432.

2376. Saur, R. L., "Corrosion testing - protective and decorative coatings", Conference on Properties of Electrodeposits, 1974, p. 170-86; Publ. by Electrochemical Society, 1975, EEA79-3733.

2377. Sinha, A. K., "Material interaction problems in semiconductor devices & integrated circuits", J. Inst. Electron. & Telecommun. Eng., 23 (9), 575-6 (1977), EEA81-19397.

2378. Smith, T., "Mechanisms of surface degradation of aluminum after the standard FPL etch for adhesive bonding", Appl. Polym. Symp., 32, 11-36 (1977), CA88-195788.

2379. Stroehle, D., "On the penetration of gases and water vapour into packages with cavities and on maximum allowable leak rates", 15th Annual Proceedings Reliability Physics Symposium, 1977, p. 101-6, EEA80-40326.

2380. Tsunashima, E., "Sandwich coating between conductive layers for the prevention of silver-migrations on a phenolic board", IEEE Trans. Components, Hybrids & Manuf. Technol., CHMT-1 (2), 182-6 (1978).

2381. Weirick, L. J., "Prevention of liquid-metal embrittlement and stress-corrosion cracking in Kovar leads", Solid State Technol., 19 (6), 55-61 (1976), EEA79-37697.

2382. Weirick, L. J., "A metallurgical analysis of stress-corrosion cracking of kovar package leads", Electrochemical Society Spring Meeting, Extended Abstracts, 1977, p. 130-2, EEA81-6106.

2383. Weissert, H., "A stable metallisation system for high reliability IC multi-chip packages", NTG-Fachber., no. 60, 180-4 (1977), German, EEA81-23644.

2384. Willmott, D. B., "Investigation of metalization failures of glass sealed ceramic dual inline integrated circuits", 15th Annual Proceedings Reliability Physics Symposium, 1977, p. 158-62, EEA80-41428.

## 4.   Encapsulation Failure

2385. Baker, H. R. and Bolster, R. N., "Moisture displacement and the prevention of surface electrical leakage", IEEE Trans. Electr. Insul., EI-11 (3), 81-5 (1976), EEA79-41626.

2386. Berg, H. M. and Paulson, W. M., "The factors affecting aluminum corrosion in plastic semiconductor packages", Electrochemical Society Spring Meeting, Extended Abstracts, 1977, p. 33-4, EEA81-4779.

2387. Cavanagh, P. J., "State-of-the-art plastic reliability", Proceedings of the Symposium on Plastic Encapsulated Semiconductor Devices, 1976, p. 10/1-11, EEA80-6690.

2388. Chaplin, N. J. and Masessa, A. J., "Reliability of epoxy and silicone molded tape-carrier silicon integrated circuits with various chip-protective coatings", 16th Annual Proceedings Reliability Physics Symposium, 1978, p. 187-93.

2389. Christou, A., "Reliability assessment of hybrid encapsulants: fluorinated network polymeric materials and silicones", Proceedings of the 13th Electrical/Electronics Insulation Conference, 1977, p. 65-9, EEA81-14350.

2390. Christou, A., Griffith, J. R. and Wilkins, B. R., "Reliability of hybrid encapsulation based on fluorinated polymeric materials", IEEE Trans. Electron Devices, ED-26 (1), 77-82 (1979).

2391. Dale, J. R. and Oldfield, R. C., "Mechanical stress likely to be encountered in the manufacture and use of plastically encapsulated devices", Microelectron. & Reliab., 16 (3), 255-8 (1977).

2392. DerMarderosian, A. and Gionet, V., "Water vapor penetration rate into enclosures with known air leak rates", 16th Annual Proceedings Reliability Physics Symposium, 1978, p. 179-86, EEA81-48758.

2393. Dubey, G. C. and Singh, R. A., "Decapsulation of epoxy potted electronic assemblies for failure analysis", J. Inst. Electron. Telecommun. Eng., 21 (9), 499-500 (1975).

2394. Ertel, A. and Perlstein, H. J., "The hermeticity hoax", 1978 Proceedings Annual Reliability and Maintainability Symposium, p. 448-53, EEA81-33014.

2395. Fitch, W. T., "Extended temperature cycling of plastic and ceramic I/Cs with thermal shock preconditioning", 14th Annual Proceedings Reliability Physics Symposium, 1976, p. 240-7.

2396. Gregoritsch, A. J., "Polyimide passivation reliability study", 14th Annual Proceedings Reliability Physics Symposium, 1976, p. 228-33, EEA80-17277.

2397. Hakim, E. B., "A case history: procurement of quality plastic encapsulated semiconductors", Solid State Technol., 20 (9), 71-3 (1977), EEA81-14685.

2398. Harrison, J. C., "Long term reliability of PED through control of the encapsulation material", Proceedings of the Symposium on Plastic Encapsulated Semiconductor Devices, 1976, Paper 9/1-25, EEA80-6609.

2399. Harrison, J. C., "Control of the encapsulation material as an aid to long term reliability in plastic encapsulated semiconductor components (PEDs)", Microelectron. & Reliab., 16 (3), 233-44 (1977), EEA81-134.

2400. Johnson, G. M. and Conaway, L. K., "Reliability evaluation of hermetic dual in-line flat microcircuit packages", Report N78-19395/0SL, McDonnell-Douglas Astronautics Co., St. Louis, Mo., 1977, 195 pp.

2401. Jones, R. O., "Developments likely to improve the reliability of plastic encapsulated devices", Microelectron. & Reliab., 17 (2), 273-8 (1978), EEA81-28440.

2402. Khajezadeh, H., "High-reliability plastic package for integrated circuits", Microelectronics, 5 (2), 28-40 (1973), EEA79-8968.

2403. Khajezadeh, H. and Rose, A. S., "Reliability evaluation of hermetic integrated circuit chips in plastic packages", 13th Annual Proceedings of Reliability Physics Symposium, 1975, p. 87-92, EEA79-46387.

2404. Khajezadeh, H. and Rose, A. S., "Reliability evaluation of trimetal integrated circuits in plastic packages", 15th Annual Proceedings Reliability Physics Symposium, 1977, p. 244-9, EEA80-41554.

2405. King, G. R., "Quality control and screening in the production of plastic encapsulated semiconductor devices (PEDs)", Micro-electron. & Reliab., 16 (3), 245-9 (1977).

2406. Koshiba, S., Yamashita, S. and Kanol, K., "Reliability to moisture-proof characteristics of resin-coated electronic parts", Bull. Electrotech. Lab., 40 (3), 230-8 (1976), EEA79-33101.

2407. Kovac, M. G., Chleck, D. and Goodman, P., "New moisture sensor for in-situ monitoring of sealed packages", Solid State Technol., 21 (2), 35-9, 53 (1978).

2408. Lowry, R. K., Van Leeuwen, C. J., Kennimer, B. L. and Miller, L. A., "Reliable dry ceramic dual in-line package (CERDIP)", 16th Annual Proceedings Reliability Physics Symposium, 1978, p. 207-12, EEA81-48760.

2409. Lycoudes, N., "The reliability of plastic microcircuits in moist environments", Solid State Technol., 21 (10), 53-62 (1978).

2410. Mann, J. E., Anderson, W. E., Raab, T. J. and Rollins, J. S., "Reliability of deposited glass", Report AD-A024664/5SL, Rockwell International, Anaheim, Calif., 1976, 136 pp.

2411. Messenger, C. G., "Improved reliability through dry and her-metic microcircuit packaging", 27th Electronic Components Conference, 1977, p. 172-4, EEA80-35964.

2412. Neighbour, F. and White, B. R., "Factors governing aluminium interconnection corrosion in plastic encapsulated microelec-tronic devices", Microelectron. & Reliab., 16 (2), 161-4 (1977), EEA80-35045.

2413. Price, S. J., Johnson, R. L. and Chapman, J. F., "For reli-able, severe environment performance, encapsulate LEDs with a clear silicone", Insul./Circuits, 23 (11), 53-6 (1977), EEA81-13305.

2414. Prince, J. L., "Investigation of factors involved in reli-ability assurance of plastic-encapsulated integrated cir-cuits", Report AD-A050875/4SL, Clemson Univ., S. C., 1978, 44 pp.

2415. Ramy, J. P., Garnier, Y. and Coudrin, J. R., "Reliability considerations on epoxy, eutectic and preform attachment of semiconductor dice", DVS Ber., 40, 133-6 (1976), CA86-49762.

2416. Reich, B., "Bias influence on corrosion of plastic encapsu-lated device metal systems", IEEE Trans. Reliab., R-25 (5), 296-8 (1976), EEA80-12540.

2417. Reich, B., "A re-examination of the bias influence on corro-sion of plastic encapsulated device (PED) metal systems", Proceedings of the Symposium on Plastic Encapsulated Semicon-ductor Devices, 1976, p. 4/1-7, EEA80-6605.

2418. Reich, B., "Reliability of plastic encapsulated semiconductor devices and integrated circuits", Solid State Technol., 21 (9), 82-8 (1978).

2419. Reich, B. and Hakim, E. B., "The use of reliable plastic semi-
      conductors in military equipment", Microelectron. & Reliab.,
      15 (1), 29-33 (1976), EEA79-28316.
2420. Ritchie, K., "Application of thermal analysis to the semi-
      conductor plastic package thermal shock testing reliability",
      Microelectron. & Reliab., 15 (5), 489-90 (1976).
2421. Roberts, B. C., "Failure analysis of plastic-encapsulated
      semiconductor devices", 1st International Conference on Plas-
      tics in Telecommunication, 1974, Paper 7, 7 pp.
2422. Ruggles, T., "Plastic power transistor reliability", Electron,
      no. 81, 54, 57 (1975), EEA79-4677.
2423. Ruthberg, S., "Gas infusion into doubled hermetic enclosures",
      IEEE Trans. Parts, Hybrids & Packag., PHP-13 (2), 110-16
      (1977), EEA80-31395.
2424. Rydin, A., "Reliability of plastic encapsulated hybrid cir-
      cuits", New Electron., 10 (6), 83-4, 87-8 (1972), EEA80-35936.
2425. Shabde, S., Edwards, J. and Meuli, W., "Moisture induced
      failure mode in a plastic encapsulated dynamic timing circuit",
      15th Annual Proceedings Reliability Physics Symposium, 1977,
      p. 33-6, EEA80-41595.
2426. Shumway, H. and Hayden, G., "Plastic power reliability", New
      Electron., 9 (23), 17, 21, 24 (1976), EEA80-2561.
2427. Somerville, D. T., "The role of hybrid construction techniques
      on sealed moisture levels", 15th Annual Proceedings Reliability
      Physics Symposium, 1977, p. 107-11, EEA80-41116.
2428. Sweet, M. E., "Assessment of moisture permeation of an epoxy
      sealed test module", Proceedings of the 28th Electronic Com-
      ponents Conference, 1978, p. 33-7, EEA81-37900.
2429. Tanino, K., Kanou, K., Kai, S. and Yanagisawa, S., "Reliability
      to moisture-proof characteristics of Japanese lacquer film.
      I.", Bull. Electrotech. Lab., 39 (11), 73-7 (1975), Japanese,
      EEA79-11896.
2430. Taylor, C. H., "Just how reliable are plastic encapsulated
      semiconductors for military applications and how can the
      maximum reliability be obtained", Microelectron. & Reliab.,
      15 (2), 131-4 (1976), EEA79-31843.
2431. Watanabe, E. and Serizawa, R., "Thermally stimulated current
      on epoxy resin coating films for electronic packaging", Mem.
      Fac. Technol. Tokyo Metrop. Univ., no. 26, 2403-11 (1976),
      EEA80-35003.
2432. Weigand, B. L., Licari, J. J. and Pratt, I. H., "The reli-
      ability of polymeric adhesives in hybrid microcircuits",
      Proc. Int. Microelectron. Symp., 1975, p. 47-53, CA86-198815.
2433. Wilson, D. D., "Decapsulation of epoxy devices using oxygen
      plasma", 15th Annual Proceedings Reliability Physics Sympos-
      ium, 1977, p. 82-4, EEA80-40325.

## 5.   Contamination and Cleaning

2434. Ameen, J. G., Rivenburgh, D. L. and Turner, J. F., "Metallo-
      graphic cleaning solution", IBM Tech. Disclosure Bull., 20
      (8), 3021 (1978).
2435. Anon., "Electrochemical removal of plating pollutants", Prod.
      Finish, 42 (7), 52-4 (1978).
2436. Anon., "Bubble cleaning for densely packed assemblies", Cir-
      cuits Manuf., 18 (9), 38 (1978).
2437. Antonoplis, R. G., "Techniques for post-PIND-test-examination
      of particle contamination in semiconductor devices", AFTA 77,
      1977, p. 56-9, EEA81-13285.
2438. Barry, K. J., "The detection and isolation of sources of ionic
      contamination on printed-wiring assemblies", AFTA 77, 1977,
      p. 38-41, EEA81-13284.
2439. Baun, W. L. and Solomon, J. S., "Surface characterization of
      contamination on adhesive bonding materials", Proc. - Int.
      Symp. Contam. Control, 4th, 1978, p. 275-8, CA89-186437.
2440. Carr, T. W., Corl, E. A., Liu, C. C. and Majtenyi, C. G.,
      "Quantitative $H_2O$ determination in components using a plasma
      chromatograph-mass spectrometer", Reliability Physics Sym-
      posium, 16th Annual Proceedings, 1978, p. 59-63, EEA81-48753.
2441. Caruso, S. V., "Some aspects of contamination detection,
      analyses, and control in microcircuits for the NASA space
      shuttle program", Proceedings of the 28th Electronic Com-
      ponents Conference, 1978, p. 28-32, EEA81-37922.
2442. Christou, A. and Wilkins, W., "Reliability aspects of moisture
      and ionic contamination diffusion through hybrid encapsulants",
      Electrochemical Society Spring Meeting, Extended Abstracts,
      1977, p. 46-7, EEA81-4780.
2443. Clementson, J. J., "Cleaning and drying with Arklone W",
      Finommech.-Mikrotech., 16 (9), 274-6 (1977), Hungarian,
      EEA81-18348.
2444. Creter, P. G. and Peters, D. E., "A new cleaning method for
      ceramic microelectronic substrates", Proceedings of the 1977
      International Microelectronics Symposium, p. 281-6.
2445. Danforth, M. A. and Sunderland, R. J., "Contamination of ad-
      hesive bonding surface treatment", Appl. Polym. Symp., no.
      32: Durability of Adhes. Bonded Struct., 1977, p. 201-15.
2446. Eichel, S. and Pettinger, F. R., "Particulate contamination
      its detection and control", Proc. of the Tech. Program, Int.
      Microelectron. Conf., 1977, p. 193-202.
2447. Elliott, D., "Contamination in solder baths and how to mini-
      mize it", Circuits Manuf., 17 (8), 42, 44-5 (1977), EEA81-19128.
2448. Elliott, D. A. and Down, W. H., "Modern pollution-free clean-
      ing methods for printed wiring assemblies", International
      Electrical, Electronics Conference and Exposition, 1977,
      p. 52-3, EEA81-23599.

2449. Engelland, G. J. and Kenyon, W. G., "Non-ionic surface con-
      tamination degrades insulation resistance", Circuits Manuf.,
      17 (6), 34-40 (1977), EEA81-14257.
2450. Ferris-Prabhu, A. V., "Time dependence of environmental-
      contaminant-induced reactions in encapsulated metallurgy",
      J. Appl. Phys., 47 (9), 4078-81 (1976), EEA80-47.
2451. Ferris-Prabhu, A. V., "Time dependence of environmental-
      contaminant-induced reactions in encapsulated metallurgy. II.
      Analysis and application of the model", J. Appl. Phys., 47
      (9), 4082-5 (1976), EEA80-48.
2452. Ferris-Prabhu, A. V., "Contaminant induced aging in inte-
      grated circuit packages", Proceedings of the 28th Electronic
      Components Conference, 1978, p. 1-6, EEA81-37899.
2453. Fisher, H. D., "Analysis of volatile contaminants in micro-
      circuits", Solid State Technol., 21 (6), 68-70, 82 (1978),
      EEA81-45188.
2454. Giger, H., "Freon cleaning of printed circuits after solder-
      ing", Finommech.-Mikrotech., 16 (6), 189-92 (1977), Hungarian,
      EEA81-10019.
2455. Jellison, J. L., "Effect of surface contamination on the
      thermocompression bondability of gold", IEEE Trans. Parts,
      Hybrids & Packag., PHP-11 (3), 206-11 (1975).
2456. Jellison, J. L., "Kinetics of thermocompression bonding to
      organic contaminated gold surfaces", IEEE Trans. Parts,
      Hybrids & Packag., PHP-13 (2), 132-7 (1977), EEA80-32253.
2457. Kenyon, W. G., "Water cleaning assemblies. I. Wave of the
      future or washout?", Insul./Circuits, 24 (2), 53-7 (1978),
      EEA81-28163.
2458. Kenyon, W. G., "PCB techniques. II. Vapor defluxing systems
      that meet today's PCB cleaning needs", Insul./Circuits, 24
      (3), 35-40 (1978), EEA81-33002.
2459. Lo, J. C. and Macur, G. J., "Ultrasonic chemical method of
      cleaning blind holes in a printed-circuit board", IBM Tech.
      Disclosure Bull., 20 (3), 962 (1977), EEA81-19141.
2460. McGuire, G. E., Jones, J. V. and Dowell, H. J., "The Auger
      analysis of contaminants that influence the thermocompression
      bonding of gold", Thin Solid Films, 45 (1), 59-68 (1977),
      EEA81-921.
2461. Reed, D., "Determining resist contamination of plating baths",
      Insul./Circuits, 23 (11), 31-3 (1977).
2462. Sbar, N. L., "Bias-humidity performance of encapsulated and
      unencapsulated Ti-Pd-Au thin-film conductors in an environment
      contaminated with $Cl_2$", IEEE Trans. Parts, Hybrids & Packag.,
      PHP-12 (13), 176-81 (1976), EEA80-3297.
2463. Sbar, N. L. and Feinstein, L. G., "Performance of new copper
      based metallization systems in an 85° 80-percent RH $Cl_2$ con-
      taminated environment", IEEE Trans. Parts, Hybrids & Packag.,
      PHP-13 (3), 208-18 (1977), EEA81-5616.

2464. Schlegel, H. and Nowotny, T., "Cleaning and passivating paste
      for silver", Patent German 2539956, Publ. March 1977,
      CA87-10152.
2465. Stanley, K. W. and Dryden, M. H., "Cleaning process for
      assembled printed-wiring boards", Finommech.-Mikrotech., 16
      (9), 263-73 (1977), Hungarian, EEA81-19134.
2466. Tatuscher, C. J., "PCB contamination introduced by the solder-
      ing process", Insul./Circuits, 24 (8), 8-18 (1978), EEA82-621.
2467. Wargotz, W. B., "Cleanliness and reliability: quantifying
      contaminant effects", Circuits Manuf., 18 (2), 40, 42, 49-9
      (1978), EEA81-49492.

## 6.  Others

2468. Agarwala, B. N., Digiacomo, G. and Joseph, R. R., "Electro-
      migration damage in aluminium-copper films", Thin Solid Films,
      34 (1), 165-9 (1976), EEA79-28619.
2469. Ager, D. J. and Mylotte, P. S., "Noise injection as a measure
      of semiconductor component performance and degradation",
      Microelectron. & Reliab., 16 (6), 679-87 (1977).
2470. Alexander, D. R., "Electrical overstress failure modeling for
      bipolar semiconductor components", IEEE Trans. Components,
      Hybrids & Manuf. Technol., CHMT-1 (4), 345-53 (1978).
2471. Anon., "How to live with silver tarnishing/silver migration",
      Circuits Manuf., 16 (10), 56, 58-9, 62 (1976), EEA80-3451.
2472. Anslow, J. W., Baxter, G. K. and Brouillette, J., "Thermal
      resistance of microelectronic packages", Report AD-A047110/
      2SL, General Electric Co., Syracuse, N. Y., 1977, 107 pp.
2473. Balenovich, J. D. and Karstaedt, W. H., "Induction heating
      effects on semiconductor disc packages", Conference Record
      of the IAS Annual Meeting, 1975, p. 340-2, EEA79-41542.
2474. Bendz, D. J., "Potentiometric test", 26th Electronic Compo-
      nents Conference, 1976, p. 168-72, EEA79-36686.
2475. Benjamin, A. G. and Pilczak, K. J., "Ruggedizing a temperature
      stable hybrid oscillator", Proceedings of the 1977 Interna-
      tional Microelectronics Symposium, p. 287-90.
2476. Berger, R. A. and Azarewicz, J. L., "Packaging effects on
      transistor radiation response", IEEE Trans. Nucl. Sci.,
      NS-22 (6), 2568-72 (1975), EEA79-28551.
2477. Bertin, A. P. and Terwilliger, T. W., "Repairs to complex
      hybrid circuits - their effect on reliability", 13th Annual
      Proceedings of Reliability Physics Symposium, 1975, p. 242-7,
      EEA79-46578.
2478. Blech, I. A., Fraser, D. B. and Haszko, S. E., "Optimization
      of Al step coverage through computer simulation and scanning
      electron microscopy", J. Vac. Sci. & Technol., 15 (1), 13-19
      (1978).
2479. Bora, J. S., "Electrical ratings of electronic components
      from the view-point of reliability", QR J., 4 (1), 21-4 (1977).

2480. Carter, W. C. and Ellozy, H. A., "Modifying circuits with lurking faults to make the circuits fault secure", IBM Tech. Disclosure Bull., 21 (6), 2665-6 (1978).

2481. Chen, C. L., "Distance 4 error correcting code that detects package-oriented failures", IBM Tech. Disclosure Bull., 20 (7), 2763 (1977).

2482. Dassatti, A. and Morelli, G., "Survey of temperature measurements for the examination of circuit reliability using an infrared detector", Telecommunicazioni, no. 62, 3-12 (1977), Italian, EEA81-16780.

2483. Davidson, J. L., Gibson, J. D., Harris, S. A. and Rossiter, T. J., "Fusing mechanism of nichrome thin films", 14th Annual Proceedings Reliability Physics Symposium, 1976, p. 173-81, CCA12-15540.

2484. Duva, R., "Gold plating defects and their underlying causes", Electron. Packag. & Prod., 18 (7), 112-14 (1978).

2485. Farrand, W. A., "Case for multichip LSI packaging of high-reliability military electronics", IEEE Trans. Parts, Hybrids & Packag., PHP-12 (4), 288-92 (1976), EEA80-13286.

2486. Frank, R. and McTigue, L., "Storage survival study reveals ways to prolong IC shelf life", Insul./Circuits, 23 (9), 32-6 (1977), EEA80-41525.

2487. Galloway, K. F. and Buehler, M. G., "The application of test structures and test patterns to the development of radiation hardened integrated circuits: a review", Report PB-256318/7SL, NBS, Washington, D. C., 1976, 15 pp.

2488. Garside, R., "The choice and reliability of electronic systems in hazardous atmospheres", Electr. Equip., 16 (10), 23, 26, 28, 33 (1977), EEA81-13341.

2489. Getten, J. R., "Plated through holes (PTH) reliability using tin bismuth solder", IBM Tech. Disclosure Bull., 20 (11), 4306 (1978).

2490. Gregoritsch, A. J., "Double level metallurgy defect study", 16th Annual Proceedings Reliability Physics Symposium, 1978, p. 27-32.

2491. Guidici, D. C., "Simple techniques of hybrid failure analysis", Electron. Packag. & Prod., 16 (5), 101-2, 104, 106 (1975), EEA79-41607.

2492. Harper, C. A., "Considerations for optimizing reliability, repairability and life cycle costs in coated circuit boards", IEEE Proc. Natl. Aerosp. Electron. Conf. NAECON '78, 1978, p. 152-7.

2493. Johnson, D. R., Knutson, R. E. and Grissom, J. T., "Testing of thick film technology in ionizing radiation environments", 27th Electronic Components Conference, 1977, p. 524-31, EEA80-35927.

2494. Kirk, W. J., Jr., Carter, L. S. and Waddell, M. L., "Eliminate static damage to circuits", Electron. Des., 24 (7), 80-5 (1976), EEA79-28313.

2495. LaCombe, D. J. and Naster, R. J., "Reliability study of micro-
      wave transistor packaging", Report AD-A020739/9SL, General
      Electric Co., Syracuse, N. Y., 1975, 162 pp.
2496. Lal, K., "Component reliability exposed to thermal neutron
      environment. III.", Microelectron. & Reliab., 16 (6), 675-7
      (1977), EEA81-9283.
2497. Lal, K., "Probability of displacement damage in a component
      exposed to nuclear radiation stress from the viewpoint of
      reliability", Microelectron. & Reliab., 17 (4), 435-9 (1978),
      EEA81-41431.
2498. Lesley, A. M., "Higher reliability for hybrids and printed
      circuit boards by functional thermal testing", IEEE Trans.
      Instrum. and Meas., IM-26 (3), 207-10 (1977), EEA80-43266.
2499. McAteer, O. J., "Electrostatic damage in hybrid assemblies",
      1978 Proceedings Annual Reliability and Maintainability
      Symposium, p. 434-42, EEA81-33032.
2500. McBrayer, J. D., Palmer, D. W. and Hickam, C. R., "Semicon-
      ductors in 300°C hybrid microcircuits", Proceedings of the
      1977 International Microelectronics Symposium, p. 241-8.
2501. Madzy, T. M., "FET circuit destruction caused by electro-
      static discharge", IEEE Trans. Electron. Devices, ED-23 (9),
      1099-1103 (1976).
2502. Mann, J. E., "Failure analysis of passive devices", 16th
      Annual Proceedings Reliability Physics Symposium, 1978,
      p. 89-92.
2503. Mann, T., "Overvoltage protection for semiconductors and
      integrated circuits", Electron, no. 88, 12, 14 (1976),
      EEA79-20532.
2504. Marques, A. M. and Partridge, J., "Progress report on nichrome
      link from reliability studies", 14th Annual Proceedings Reli-
      ability Physics Symposium, 1976, p. 182-91.
2505. May, T. C. and Woods, M. H., "Alpha-particle-induced soft
      errors in dynamic memories", IEEE Trans. Electron. Devices,
      ED-26 (1), 2-9 (1979).
2506. Minear, R. L. and Dodson, G. A., "Effects of electrostatic
      discharge on linear bipolar integrated circuits", 15th Annual
      Proceedings Reliability Physics Symposium, 1977, p. 138-43.
2507. Narayanan, M. and Soni, V., "Application of fluorinert liquid
      in the failure analysis of semiconductor integrated circuits",
      Solid State Technol., 21 (2), 45-7 (1978).
2508. Palmer, D. W., "Hybrid microcircuitry for 300° operation",
      IEEE Trans. Parts, Hybrids & Packag., PHP-13 (3), 252-7
      (1977), EEA81-909.
2509. Palmer, D. W. and Heckman, R. C., "Extreme temperature range
      microelectronics", IEEE Trans. Components, Hybrids & Manuf.
      Technol., CHMT-1 (4), 333-40 (1978).
2510. Sakamoto, Y., Kondo, Y., Ejiri, M., Takagi, T., Kawase, H.
      and Oshio, T., "The consideration of reliability in a fully
      automatic assembly system of small signal transistors", 27th
      Electronic Components Conference, 1977, p. 146-53, EEA80-35963.

2511. Smith, J. S., "Electrical overstress failure analysis in micro-circuits", 16th Annual Proceedings Reliability Physics Symposium, 1978, p. 41-6.

2512. Speranskiy, D. V., "A method for construction of hybrid chips which can be diagnosed with a specified resolution capability", Eng. Cybern., 15 (4), 118-25 (1977), EEA81-42159.

2513. Thomas, R. W., "Moisture, myths, and microcircuits", IEEE Trans. Parts, Hybrids & Packag., PHP-12 (13), 167-71 (1976), EEA80-45.

2514. Vest, R. W., "The effects of substrate composition on thick film circuit reliability", Report AD-A056076/3SL, Purdue Research Foundation, Lafayette, Ind., 1978, 24 pp.

2515. Villella, F. and Nowakowski, M. F., "Device/packaging reliability problems of solid-state devices for space applications", EASCON 76 Record, 1976, Paper 145A, 9 pp., EEA80-2562.

2516. Walker, R. C. and Rickers, H. C., "Semiconductor electrostatic discharge damage protection", SAE Prepr. No. 770228, 1977, 12 pp.

2517. Woods, M. H. and Gear, G., "A new electrostatic discharge failure mode", IEEE Trans. Electron. Devices, ED-26 (1), 16-21 (1979).

2518. Wright, W. L., "Electrostatic discharge protection of FET gates with thin film capacitors", IBM Tech. Disclosure Bull., 21 (6), 2396-7 (1978).

2519. Wunsch, D. C., "Application of electrical overstress models to gate protective networks", 16th Annual Proceedings Reliability Physics Symposium, 1978, p. 47-55.

2520. Yaney, D. S., Nelson, J. T. and Vanskike, L. L., "Alpha-particle tracks in silicon and their effect on dynamic MOS RAM reliability", IEEE Trans. Electron. Devices, ED-26 (1), 10-16 (1979).

2521. Yeh, C. S., "Noise analysis and potential failure mechanisms in hybrid microcircuits and components", AFTA 77, 1977, p. 64-74, EEA81-14351.

2522. Yu, C. C., "Monitoring stud diode package degradation", IBM Tech. Disclosure Bull., 20 (2), 671 (1977), EEA81-18379.

XXIII.   TESTING
        1.   Accelerated Life Test

2523. Aitken, A. and Kung, P., "Influence of design and process parameters on the reliability of CMOS integrated circuits", Microelectron. & Reliab., 17 (1), 201-10 (1977).

2524. Anon., "Temperature cycling vs. steady-state burn-in", Circuits Manuf., 16 (9), 52-4 (1976), EEA80-8605.

2525. Anon., "A very long term test", Eltek. Aktuell Elektron. A, 19 (13), 55 (1976), Swedish, EEA80-3251.

2526. Anon., "Accelerated life testing effects on CMOS microcircuit characteristics", Report N77-31406/0SL, RCA Solid State Div., Somerville, N. J., 1977, 54 pp.

2527. Bailey, C. M., "Effects of burn-in and temperature cycling on the corrosion resistance of plastic encapsulated integrated circuits", 15th Annual Proceedings Reliability Physics Symposium, 1977, p. 120-4, EEA80-41570.

2528. Bartels, D., Capobianco, A., Lasch, K. B. and Lynch, K. S., "Reliability evaluation of ECL integrated circuits", 15th Annual Proceedings Reliability Physics Symposium, 1977, p. 196-203.

2529. Baxter, G. K., "Recommendation of thermal measurement techniques for IC chips and packages", 15th Annual Proceedings Reliability Physics Symposium, 1977, p. 204-11.

2530. Bratschun, W. R., Everett, P. G. and Gabrykewicz, T. J., "The reliability of thick film capacitors and crossovers", 27th Electronic Components Conference, 1977, p. 161-8, EEA80-35920.

2531. Carpenter, M. R. and Fitch, W., "Thermomechanical testing techniques for microcircuits", Report AD-A014345/3SL, Motorola Inc., Phoenix, Ariz., 1975, 265 pp.

2532. Der Marderosian, A. and Murphy, C., "Humidity threshold variations for dendrite growth on hybrid substrates", 15th Annual Proceedings Reliability Physics Symposium, 1977, p. 92-100, EEA80-41115.

2533. DiNitto, J. R., Lasch, K. B. and Farrell, J. P., "Prelid burn-in of hybrid circuits", Proceedings of the 28th Electronic Components Conference, 1978, p. 340-3, EEA81-42161.

2534. Edward, O. and Lycoudes, N., "EPIIC: environmental package indicators for integrated circuits", Electron. Packag. & Prod., 15 (10), 103-4, 106 (1975).

2535. Ferro, J. A., "An accelerated method for effective process control of plastic encapsulated nichrome PROMs", 15th Annual Proceedings Reliability Physics Symposium, 1977, p. 125-7, EEA80-41552.

2536. Fitch, W. T., "The degradation of bonding wires and sealing glasses with extended thermal cycling", 13th Annual Proceedings of Reliability Physics Symposium, 1976, p. 58-69, EEA79-46386.

2537. Fitch, W. T., "Extended temperature cycling of plastic and ceramic I/C's with thermal shock preconditioning", 14th Annual Proceedings Reliability Physics Symposium, 1976, p. 240-7, EEA80-17032.

2538. Fort, E. M. and Pietsch, H. E., "Aging of insulation by thermal and electrical stresses", 12th Electrical/Electronics Insulation Conference, 1975, p. 143-6, EEA79-46583.

2539. Gallace, L. J., Khajezadeh, H. J. and Rose, A. S., "Accelerated reliability evaluation of trimetal integrated circuit chips in plastic packages", 16th Annual Proceedings Reliability Physics Symposium, 1978, p. 224-8.

2540. Goedbloed, W., Hieber, H. and van Nie, A. G., "Ageing tests on microwave integrated circuits", Radio Electron. Eng., 48 (1-2), 13-22 (1978).

2541. Hall, P. M., Panousis, N. T. and Menzel, P. R., "Strength of gold-plated copper leads on thin film circuits under accelerated aging", IEEE Trans. Parts, Hybrids & Packag., PHP-11 (3), 202-5 (1975).

2542. Harada, T. and Shiomi, H., "Accelerated life test of aluminium electrolytic capacitors and their acceleration factors", Bull. Electrotech. Lab., 40 (9), 763-83 (1976), Japanese, EEA80-3250.

2543. Hieber, H., Betke, F. and Pape, K., "Ageing tests on gold layers and bonded contacts", Electrocompon. Sci. & Technol., 4 (2), 89-94 (1977), EEA81-5631.

2544. Hirsch, L., "Accelerated electrical testing for improved device reliability", 15th Annual Proceedings Reliability Physics Symposium, 1977, p. 255-6.

2545. Johnson, G. M., "Accelerated testing highlights CMOS failure modes", EASCON 76 Record, 1976, p. 142A/1-9, EEA80-3554.

2546. Johnson, G. M. and Stitch, M., "Microcircuit accelerated testing reveals life limiting failure modes", 15th Annual Proceedings Reliability Physics Symposium, 1977, p. 179-95, EEA80-41553.

2547. Kalmar, G. and Balog, B., "Reliability test system for TTL integrated circuits with DIL package", 3rd Symposium on the Reliability of Electronics, 1973, p. 153-69, Russian, CCA9-6184.

2548. Kang, S. K., Zommer, N. D., Feucht, D. L. and Heckel, R. W., "Thermal fatigue failure of soft-soldered contacts to silicon power transistors", IEEE Trans. Parts, Hybrids & Packag., PHP-13 (3), 318-21 (1977).

2549. Khadpe, S., "Aging characteristics of aluminum wire bonds on thick film platinum-silver metalization", Insul./Circuits, 23 (2), 23-4 (1977), EEA80-13283.

2550. Khadpe, S., "Corrosion characteristics of passivated and unpassivated Al-to-PtAg wire bonds under accelerated temperature-humidity conditions", Proceedings of the 13th Electrical/Electronics Insulation Conference, 1977, p. 61-4, EEA81-14361.

2551. Kiessling, K. G., "Accelerated reliability tests on electronic components and equipment", Radio Fernsehen Elektron., 26 (2), 43-4 (1977), German, EEA80-28012.

2552. Kobayashi, T. and Ariyoshi, H., "Reliability of hybrid integrated circuits for 20 G-400 M system", Rev. Electr. Commun. Lab., 24 (7-8), 632-41 (1976), EEA79-48295.

2553. Koelmans, H. and Kretschman, H. J., "Water droplet formation during the life testing of IC's in a humid ambient", J. Electrochem. Soc., 125 (10), 1715-16 (1978).

2554. Lesley, A. M., "Higher reliability for hybrids and printed circuit boards by functional thermal testing", Autotestcon '76, 1976, p. 70-5, EEA80-10866.

2555. Lethonen, D. E., "Microcircuit reliability assessment through accelerated testing", Electron. Packag. & Prod., 17 (7), 288-93 (1977).

2556. Macinlek, R. B., Pisacich, E. D. and Speerschneider, C. J., "Thermal aging characteristics of In-PB solder bonds to gold", Proceedings of the 1977 International Microelectronics Symposium, p. 209-12.

2557. Marquardt, R. A., "Five-year life test data on pressure (10 000 lbf/in$^2$) tolerant electronic components", IEEE Trans. Components, Hybrids & Manuf. Technol., CHMT-1 (4), 365-71 (1978).

2558. Maximow, B., Reiss, E. M. and Kukunaris, S., "Accelerated testing of class A CMOS integrated circuits", 15th Annual Proceedings Reliability Physics Symposium, 1977, p. 212-16.

2559. Nakada, Y. and Schock, T. L., "Reliability study of a high-precision thick film resistor network", IEEE Trans. Parts, Hybrids & Packag., PHP-13 (3), 229-34 (1977).

2560. Nakazawa, S., Shimada, T. and Shiomi, H., "Life-time acceleration for tantalum resistor under the humid environment", Bull. Electrotech. Lab., 41 (7), 556-61 (1977), Japanese, EEA81-5528.

2561. Peeples, J. W., "Electrical bias level influence on THB results of plastic encapsulated NMOS 4K RAM's", IEEE Trans. Electron. Devices, ED-26 (1), 72-6 (1979).

2562. Reynolds, F. H., "Accelerated-test procedures for semiconductor components", 15th Annual Proceedings Reliability Physics Symposium, 1977, p. 166-78.

2563. Sbar, N. L. and Kozakiewicz, R. P., "New acceleration factors for temperature, humidity, bias testing", IEEE Trans. Electron. Devices, ED-26 (1), 56-71 (1979).

2564. Schnable, G. I., Reiss, E. M. and Vincoff, M., "Reliability of hermetically-sealed CMOS integrated circuits", IEEE Electronics and Aerospace Systems Convention (EASCON '76), Record, 1976, Paper 143, 7 pp., EEA80-3555.

2565. Schwartz, N. and Bacon, D. D., "SEAT: simulated environment accelerated test", J. Electrochem. Soc., 125 (9), 1487-93 (1978).

2566. Svitak, J. J. and Williams, A. F., "Influence of thermal aging on the electrical and mechanical properties of conductive adhesives", Proceedings of the 1977 International Microelectronics Symposium, p. 189-96.

2567. Taylour, C. H., "Reliability testing of BAC hybrid circuits", Microelectron. & Reliab., 16 (4), 295-302 (1977), EEA81-10052.

2568. Vasilev, G. K., Makarov, N. I. and Prokhorov, Y. I., "Cyclic heating of a film on a substrate", J. Eng. Phys., 28 (2), 227-31 (1975), EEA80-9392.

2569. Weissflug, V. A. and Sisul, E. V., "Cyclic and low temperature effects on microcircuits", Report N78-16264/1SL, McDonnell-Douglas Astronautics Co., St. Louis, Mo., 1977, 155 pp.

2570. Wilson, G. C., "Accelerated ageing of P.C. boards to test solderability", New Electron., 10 (6), 99-100, 103-4 (1977).

2571. Zierdt, C. H., Jr., "Accelerated life testing for LSI failure mechanisms", 16th Annual Proceedings Reliability Physics Symposium, 1978, p. 76-8.

## 2. Boards/Cards

2572. Abramson, R. F., Bailey, D. L., Reddi, U. P., Rupp, G. and
      Thompson, T. W., "Reduced fault testing for array logic board",
      IBM Tech. Disclosure Bull., 21 (7), 2678-9 (1978).
2573. Allen, D. P., "Comparative board testing reaches new highs in
      diagnostic speed", Commun. Int., 4 (2), 26-8 (1977), EEA80-2387.
2574. Anderson, R. E., "Trends in PC board testing", Electron.
      Packag. & Prod., 17 (9), Pt. 2, T6-T8, T10, T12 (1977).
2575. Beaven, P. A., Llewelyn, R. J. and Rowe, H. L., "Card testing
      a microprocessor", IBM Tech. Disclosure Bull., 21 (4), 1664-5
      (1978).
2576. Beaven, P. A. and Monro, I. D., "Microprocessor card test",
      IBM Tech. Disclosure Bull., 20 (11), 4602-5 (1978).
2577. Bottorff, P. and Muehldorf, E. I., "Impact of LSI on complex
      digital circuit board testing", 1977 Electro Conference
      Record, 1977, Paper 32/3/1-12, EEA81-13958.
2578. Chase, J., "Computer-controlled signature testing of memory
      boards", Digest of Papers 1978 Semiconductor Test Conference,
      p. 235-8.
2579. Ciaramella, A., "Chip and board digital testing", Electron.
      & Telecommun., 26 (3), 123-7 (1977), Italian, EEA81-4757.
2580. Foley, G., "Designing microprocessor boards for testability",
      Digest of Papers 1978 Semiconductor Test Conference, p. 176-80.
2581. Grason, J., "The impact of a complex board level circuit on
      automatic test aids", Digest of Papers 1978 Semiconductor
      Test Conference, p. 258-63.
2582. Hack, G. E., "Circuit module interconnection testing", IBM
      Tech. Disclosure Bull., 20 (3), 1102 (1977), EEA81-22856.
2583. Hsu, F. C., "A comparison study of the techniques for deriv-
      ing component diagnosis for manufacturing testing of cards and
      boards", Digest of Papers 1977 Semiconductor Test Symposium,
      p. 192-5, EEA81-18341.
2584. Illes, G., "Modularity in test system architecture", Digest
      of Papers 1978 Semiconductor Test Conference, p. 192-5.
2585. Jones, W. K., "The Murphy's law approach to testing subassem-
      blies", Circuits Manuf., 18 (12), 21-22, 26, 28-32, 34 (1978).
2586. Kadakia, V. and Holt, C., "Design partitioning for micropro-
      cessor board test", Proc. of the Tech. Program, Int. Micro-
      electron. Conf., 1977, p. 5-11.
2587. Lesley, A. M., "Higher reliablity for hybrids and printed
      circuit boards by functional thermal testing", IEEE Trans.
      Instrum. Meas., IM-26 (3), 207-10 (1977).
2588. Lyons, N. and Hamilton, G., "Microprocessor board testing",
      New Electron., 10 (19), 100, 103 (1977), EEA81-13281.
2589. Neuhausel, J. E., "Testing microprocessor boards", Circuits
      Manuf., 19 (1), 52, 54, 56, 61 (1979).
2590. Oberly, R. P., "How to beat the card test game (logic test-
      ing)", Digest of Papers 1977 Semiconductor Test Symposium,
      p. 16-18, EEA81-18904.

2591. Oberly, R. P. and Strenk, J. E., "More board test coverage, please", 1977 Electro Conference Record, 1977, Paper 18/3/1-3, EEA81-13333.

2592. Ostrego, M., "Functional board testing in the production environment", Digest of Papers 1978 Semiconductor Test Conference, p. 79-80.

2593. Protheroe, G., "High volume PCB testing", Electron. Prod., 7 (1), 25, 27 (1978), EEA81-37882.

2594. Purks, S. R., "Flexibility for testing boards containing LSI components", 1977 Electro Conference Record, Paper 32/2/1-5.

2595. Purks, S. R., "Experiences with ATE providing testability of microprocessor boards", IEEE Trans. Instrum. Meas., IM-27 (2), 178-81 (1978).

2596. Reis, R. J., "Circuit board input/output connector test method", IBM Tech. Disclosure Bull., 20 (12), 5258-9 (1978).

2597. Rieseler, F. H., "Digital circuit board test considerations in a rapid growth situation", Proc. Tech. Program Natl. Electron. Packag. Prod. Conf., 1977, p. 220-25.

2598. Silverstein, M., "Why a memory board/system tester?", Digest of Papers 1978 Semiconductor Test Conference, p. 228-34.

2599. Stewart, J. H., "Future testing of large LSI circuit cards", Digest of Papers 1977 Semiconductor Test Symposium, p. 6-15, EEA81-18903.

2600. Sulman, D., "Clock-rate testing of LSI circuit boards", Digest of Papers 1978 Semiconductor Test Conference, p. 66-70.

2601. Taschioglou, K., "An overview of LSI board level testing today", Digest of Papers 1978 Semiconductor Test Conference, p. 8-12.

2602. West, B. G., "LSI PC assembly testing using parallel pin electronics", Solid State Technol., 21 (10), 80-2 (1978).

### 3. Bonds

2603. Albers, J., "The destructive bond pull test", Report NBS-SP-400-18, Nat. Bur. Stand., Washington, D. C., 1976, 44 pp., EEA79-33167.

2604. Amey, D. I. and Moore, R. P., "Chip carrier/elastomer interconnecting element test results", Electr. Contacts: Annu. Holm. Conf. on Electr. Contacts, Proc., 23rd, 1977, pt. 1, p. 51-7.

2605. Babiel, J. J. and Zobniw, L. M., "Shorts diagnosis using logic test data", IBM Tech. Disclosure Bull., 21 (2), 503-8 (1978).

2606. Bascom, W. D. and Bitner, H. L., "A fracture approach to thick film adhesion measurements", J. Mater. Sci., 12 (7), 1401-10 (1977), EEA80-28626.

2607. Becher, P. F. and Mewell, W. L., "Adherence-fracture energy of a glass-bonded thick film conductor - effect of firing conditions", J. Mater. Sci., 12 (1), 90-6 (1977).

2608. Becher, P. F. and Murday, J. S., "Thick film adherence fracture energy: influence of alumina substrates", J. Mater. Sci., 12 (6), 1088-94 (1977), EEA80-28625.

2609. Bekeshko, N. A., "Investigation of the active thermal method of inspecting soldered joints", Sov. J. Nondestr. Test., 13 (4), 407-11 (1977).

2610. Brassell, G. W. and Fancher, D. R., "Rating hybrid adhesives", Circuits Manuf., 18 (10), 50, 52, 54 (1978).

2611. Cada, O. and Smela, N., "Some perceptions from long-term testing of adhesive bonding of plastics with metals", Adhaesion, 21, 79-83 (1977), German, CA87-24037.

2612. Cammarano, A. S. and Hinderer, J. J., "Optical testing of solder pads", IBM Tech. Disclosure Bull., 21 (7), 2914-15 (1978).

2613. Cohen, B., "How to use infrared microscopy to non-destructively evaluate [die attach] wire bonds [on Si chips]", Insul./Circuits, 24 (9), 13-14 (1978), EEA82-653.

2614. Coleman, M. V. and Gurnett, G. E., "The structure and properties of reactively bonded thick film gold conductors", Electrocompon. Sci. & Technol., 5 (1), 55-9 (1978), EEA81-37912.

2615. Czanderna, A. W., Vasofsky, R. and Czanderna, K. K., "Mass changes of adhesives during curing, exposure to water vapor, evacuation and outgassing", Proceedings of the 1977 International Microelectronics Symposium, p. 197-208.

2616. Deskur, K. J., "Short circuit detection", IBM Tech. Disclosure Bull., 20 (7), 2627-9 (1977), EEA81-45723.

2617. Deutsch, R. J., Longenbach, H. N. and Walbert, G. R., "Automatic data collection for pull test (thin-film circuit leads)", Tech. Dig., no. 40, 9 (1975), EEA79-8978.

2618. Dor-Ram, J., Weiss, B. Z. and Komem, Y., "Study of explosion clad bonding by transmission electron microscopy", Metall. Trans. A, 8A (3), 518-20 (1977).

2619. Drapella, A., Wierzba, H. and Wroczynski, P., "Method of the defective soldered connections detection", Elektronika, 18 (9), 343-6 (1977), Polish, EEA81-13287.

2620. Ellingson, A. C., "Inspection of soldered electrical connections", Insul./Circuits, 22 (11), 17-23 (1976), EEA80-12547.

2621. Engel, P. A., Roshon, D. D. and Thorne, D. A., "Coating bond strength test method", IBM Tech. Disclosure Bull., 20 (11), 4310 (1978).

2622. Engel, P. A., Roshon, D. D. and Thorne, D. A., "Indentation tests for plastic-to-copper bond strength", Insul./Circuits, 24 (12), 35-6 (1978).

2623. Fancher, D. R. and McDaniel, J. E., "Seal and rework evaluation of seam-welded hybrid packages", Electron. Packag. Prod., 18 (2), 88-90, 92, 94 (1978).

2624. Fouts, D. P. and Jaspal, J. S., "Solder joint test vehicle", IBM Tech. Disclosure Bull., 20 (10), 3915-16 (1978).

2625. Gerlich, V. and Joch, P., "Electrical properties of joints made with epoxy resins", Fernmeldetechnik, 18 (3), 114-16 (1978), German, EEA81-48747.

2626. Hagy, H. E., "Thermal differential tables aid in estimating seal stresses", Electron. Packag. & Prod., 18 (7), 182-6, 188, 190 (1978).

2627. Harman, G. G. and Cannon, C. A., "The microelectronic wire bond pull test - how to use it, how to abuse it", IEEE Trans. Components, Hybrids & Manuf. Technol., CHMT-1 (3), 203-10 (1978), EEA81-49509.

2628. Harzdorf, F., "Non-destructive testing of soldered joints. Assessment of joint quality", Fernmeldetechnik, 17 (1), 18-20 (1977), German, EEA80-27966.

2629. Hauer, H., "Properties of soft-soldered joints on gold plated thin film conductors", NTG-Fachber., no. 60, 185-90 (1977), German, EEA81-23628.

2630. Hitch, T. T., "Adhesion measurements on thick-film conductors", ASTM Spec. Tech. Publ. No. 640; Adhes. Meas. of Thin Films, Thick Films, and Bulk Coat., 1976, p. 211-32, Publ. 1978.

2631. Holyfield, S., "Interconnection network testing", Electron. Packag. & Prod., 15 (10), 36-8, 40, 42, 44 (1975).

2632. Johnson, D. R. and Chavez, E. L., "Characterization of the thermosonic wire bonding technique", Report SAND-76-5292, Sandia Labs., Albuqerque, N. Mex., 1976, 16 pp.

2633. Kadereit, H. G. and Schlemm, A., "Adhesion measurements on conductive layers for hybrid microcircuits", Siemens Forsch.- & Entwicklungsber., 6 (4), 220-5 (1977), German, EEA80-35937.

2634. Kadereit, H. G. and Schlemm, A., "Adhesion measurements of metallization for hybrid microcircuits", Electrocompon. Sci. & Technol., 4 (3-4), 147-60 (1977), EEA81-14336.

2635. Kohara, M., Harada, H., Sakane, H. and Michii, K., "Gang-bonding strength using AuSn eutectic reaction", Trans. Inst. Electron. & Commun. Eng. Jpn. Sect. E, E59 (10), 43 (1976), EEA80-41120.

2636. Koshy, T. C., "Non-destructive testing of adhesive bonds using Fokker bond tester", J. Inst. Eng. (India) Mech. Eng. Div., 57 (pt. ME-3), 137-9 (1976), EEA80-16644.

2637. Kossowsky, R., "Characterization of reaction bonded Au and Ag thick film metallizations", 15th Annual Proceedings Reliability Physics Symposium, 1977, p. 262-71, EEA80-41110.

2638. Koudounaris, A., "Test procedure to evaluate relative failure characteristics of wire bonding methods", Insul./Circuits, 24 (3), 43-4 (1978), EEA81-33035.

2639. Lange, Y. V., "Acoustic amplitude method of inspecting bonding in laminated structures", Sov. J. Nondestr. Test., 12 (1), 5-11 (1976).

2640. Leven, S. S., "Adhesion measurement technique for soldered thick-film conductors", ASTM Spec. Tech. Publ. No. 640; Adhes. Meas. of Thin Films, Thick Films, and Bulk Coat., 1976, p. 269-84, Publ. 1978.

2641. Mitnick, W. L., "Computer detection of bent fingers in lead bonding frames", Report 76B-00695, MIT, Cambridge, Mass., 1976, 18 pp.

2642. Morey, R. L., "Adhesion of thick films to ceramic and its measurement by both destructive and nondestructive means", ASTM Spec. Tech. Publ. No. 640; Adhes. Meas. of Thin Films, Thick Films, and Bulk Coat., 1976, p. 233-50, Publ. 1978.

2643. Peckinpaugh, C. J., "Adhesion loss in solderable thick film conductor systems as a result of thermal shock", Insul./Circuits, 23 (3), 35-8 (1977), EEA80-24617.

2644. Perkins, K. L., Licari, J. J. and Caruso, S. V., "Evaluation of adhesives for hybrid microcircuit package sealing", IEEE Trans. Components, Hybrids & Manuf. Technol., CHMT-1 (4), 412-15 (1978).

2645. Perlman, A. L., "Method for aiding visual inspection of C-4 pads", IBM Tech. Disclosure Bull., 21 (3), 1038 (1978).

2646. Piccoli, S. C. and Lavery, W., "Behavior of 1 mil aluminum wire at high temperature for hybrid microcircuits", AFTA 77, 1977, p. 94-8, EEA81-14352.

2647. Roddy, J., Spann, N. and Seese, P., "Nondestructive bond pull in high-reliability applications", IEEE Trans. Components, Hybrids & Manuf. Technol., CHMT-1 (3), 228-36 (1978), EEA81-49513.

2648. Sacher, E. and Woods, J. J., "Improving bonding strength of polyimide-to-silicon wafers", IBM Tech. Disclosure Bull., 20 (6), 2161 (1977).

2649. Scheel, W., Kurkowaski, F. and Schroder, G., "Seal-testing glass-encased components", Radio Fernsehen Elektron., 25 (5), 153-4 (1976), German, EEA79-27569.

2650. Semenov, G. M. and Demina, S. I., "Inspecting the quality of the contact joints of stacked semiconductor power devices", Sov. J. Nondestr. Test., 13 (3), 362-3 (1977), EEA81-42387.

2651. Telfer, T. A., "Method of determining bond quality of power transistors attached to substrates", Patent USA 3889122, Publ. June 1975, EEA79-4772.

2652. Unal, A. and Herrmann, G., "On the effect of bonding in a laminated composite", Report AD-A015878/2SL, Stanford Univ., Calif., 1975, 33 pp.

2653. Waser, S., "Back-bias continuity checks TTL wire bonds", Electronics, 48 (9), 98-9 (1975).

2654. Westphal, K., "Bond strength of metal layers on strip line substrates", Nachrichtentech. Elektron., 26 (2), 70-3 (1976), German, EEA79-20563.

2655. Zumberov, V. V., Plotnikov, E. A., Sokov, V. A. and Tolmachev, Y. I., "Ultrasonic method for the inspection of soldered and fused junctions in semiconductor driver units", Sov. J. Nondestr. Test., 11 (2), 244-6 (1975), EEA80-35015.

4.  Coatings

2656. Abshier, C. S., Berry, J. and Maget, H. J., "Toughening agents improve epoxy encapsulants", Insul./Circuits, 23 (11), 27-9 (1977), EEA81-13303.

2657. Andrews, E. H. and Stevenson, A., "Fracture energy of epoxy resin under plane strain conditions", J. Mater. Sci., 13 (8), 1680-8 (1978).

2658. Anon., "Quality assessment. X. Epoxy resin smear", Circuit World, 4 (2), 12-13 (1978), EEA81-37879.

2659. Armstrong, G., Greenberg, A. S. and Sites, J. R., "Very low temperature thermal conductivity and optical properties of stycast 1266 epoxy", Rev. Sci. Instrum., 49 (3), 345-7 (1978).

2660. Bagnall, R. T. and Schroter, S., "Tests and evaluation of polymers for hybrid assembly", 12th Electrical/Electronics Insulation Conference, 1975, p. 83-4, EEA79-46552.

2661. Barton, G. T. and Harriot, R. W., "Testing conductive epoxies: technique defects small but important changes in resistance", Circuits Manuf., 17 (10), 44-6 (1977).

2662. Basile, L. and Fantini, F., "Problems associated with evaluation of plastic housing for long useful life applications", Alta Freq., 45 (5), 324-34 (1976), Italian, EEA79-44923.

2663. Baxter, L. L., "Report on toxicity studies of epoxy powder hardeners", Insul./Circuits, 22 (10), 39-41 (1976), EEA80-44.

2664. Bolger, J. C., "Epoxy encapsulation compounds with improved combinations of thermal shock, heat and reversion resistance", IEEE Proceedings National Aerospace Electronics Conference, 1975, p. 101.

2665. Bucknall, C. B. and Yoshii, T., "Relationship between structure and mechanical properties in rubber-toughened epoxy resins", Br. Polym. J., 10 (1), 53-9 (1978).

2666. Buhrer, B., "Materials for moulded-plastic-clad l.v distribution systems", BBC Nachr., 58 (1), 45-50 (1976), German, EEA79-11895.

2667. Castonguay, R. N., "New fiberglass fabric reinforced plastic substrates for hybrid circuits", Proc. Int. Microelectron. Symp., 1976, p. 45-51, CA86-198813.

2668. Chaplin, N. J. and Masessa, A. J., "Evaluation of encapsulants for tape-carrier silicon IC's", Insul./Circuits, 24 (10), 39-43 (1978).

2669. Chellis, L. N., "Stabilizing aromatic polyamide fabrics", IBM Tech. Disclosure Bull., 20 (2), 555 (1977), EEA81-19138.

2670. Christou, A., "Assessment of silicone encapsulants for hybrid integrated circuits (HIC)", IEEE Trans. Parts, Hybrids & Packag., PHP-13 (3), 298-304 (1977), EEA81-5636.

2671. Christou, A. and Wilkins, W., "Assessment of silicone encapsulation materials: screening techniques", 15th Annual Proceedings Reliability Physics Symposium, 1977, p. 112-19, EEA80-40634.

2672. Coisson, R., Paracchini, C. and Schianchi, G., "Electrolumi-
      nescence in an epoxy resins", J. Electrochem. Soc., 125 (4),
      581-3 (1978).
2673. Comizzoli, R. B., "Nondestructive, reverse decoration of
      defects in IC passivation overcoats", J. Electrochem. Soc.,
      124 (7), 1087-95 (1977), EEA80-41526.
2674. Cooper, R., Varlow, B. R. and White, J. P., "Electric strength
      of prestressed polythene following sudden cooling", Proc.
      Inst. Electr. Eng., 123 (2), 187-8 (1976), EEA79-16914.
2675. Dale, J. R. and Oldfield, R. C., "Mechanical stresses likely
      to be encountered in the manufacture and use of plastically
      encapsulated devices", Microelectron. & Reliab., 16 (3),
      255-8 (1977), EEA81-137.
2676. Delmonte, J., "Urethane or epoxy? Which encapsulant to use",
      Insul./Circuits, 23 (11), 37-8 (1977).
2677. DiLeo, D. A., Svitak, J. J. and Williams, A. F., "Conductive
      epoxies for microelectronics - electrical and mechanical
      properties characterization", Proc. of the Tech. Program, Int.
      Microelectron. Conf., 1977, p. 29-36.
2678. Eberhardt, M., "Influence of the ambient temperature on the
      electrical ageing of epoxy resins and epoxy resin-mica in-
      sulations", Elektrie, 30 (1), 20-1 (1976), German, EEA79-18765.
2679. Eisenmann, D. E. and Halyard, S. M., Jr., "Thermal character-
      ization of polymeric materials used in semiconductor die
      bonding", Thermochim. Acta, 14 (1-2), 87-97 (1976), CA84-158717.
2680. El-Aasser, M. S., Vanderhoff, J. W., Misra, S. C. and Manson,
      J. A., "Electron microscopy of epoxy latexes and their films",
      J. Coat. Technol., 49 (635), 71-8 (1977).
2681. Elmore, G. V., "Surface preparation of epoxy laminate to pro-
      mote bonding of electroless copper", J. Electrochem. Soc.,
      124 (1), 89-91 (1977), EEA80-12879.
2682. Elsby, T. W., "Thermal characterization of epoxy and alloy
      attachment of hybrid components", 27th Electronic Components
      Conference, 1977, p. 320-3, EEA80-35951.
2683. Fox, M. J., "A comparison of the performance of plastic and
      ceramic encapsulations based on evaluation of CMOS integrated
      circuits", Microelectron. & Reliab., 16 (3), 251-4 (1977),
      EEA81-136.
2684. Georgi, A., "Influence of the manufacturing technique on
      the electric breakdown of epoxy-resin insulations subjected
      to a long-time alternating-current voltage stress", Elektrie,
      30 (5), 277-80 (1976), German, EEA79-33175.
2685. Gillham, J. K., Glandt, C. A. and McPherson, C. A., "Char-
      acterization of thermosetting epoxy systems using a torsional
      pendulum", Report AD-A045886/9SL, Princeton Univ., N. J.,
      Dept. of Chemical Engineering, 1977, 33 pp.
2686. Gledhill, R. A., Kinloch, A. J., Yamini, S. and Young, R. J.,
      "Relationship between mechanical properties of and crack
      propagation in epoxy resin adhesives", Polymer, 19 (5),
      574-82 (1978).

2687. Goebell, J., "Aspects of selecting encapsulating resins for cables", Schweiz. Tech. Z., no. 31–32, 757–60 (1977), German, EEA81–138.

2688. Guarino, D. A., "Rigid PVC packaging", SPE, Del Val Section, Regional Technical Conference, Technical Papers, 1977, p. 123–7.

2689. Guertin, R., "Hermeticity test pinpoints leaks in thick film hybrids without damaging components", Insul./Circuits, 23 (12), 26 (1977), EEA81–14340.

2690. Hakim, E. B., "A case history: procurement of quality plastic encapsulated semiconductors", Solid State Technol., 20 (9), 71–3 (1977).

2691. Hakim, E. B., "U.S. Army Panama field test of plastic encapsulated devices", Microelectron. & Reliab., 17 (3), 287–92 (1978), EEA81–32289.

2692. Hallworth, A., "Accurate metering of two-part resins necessary for uniform encapsulation", Can. Electron. Eng., 20 (2), 23–4 (1976), EEA79–19567.

2693. Hanley, L. D. and Martin, J. H., "A test of parylene as a protective system for microcircuitry", 13th Annual Proceedings of Reliability Physics Symposium, 1975, p. 53–7, EEA79–44931.

2694. Hickman, J. J., Regester, R. F. and Wallig, L. R., "High reliability with the dry-film solder mask system", Proc. of the Electr./Electron. Insul. Conf., 13th, 1977, p. 80–5.

2695. Isleifson, R. E. and Carlstrom, G. L., "Piezoelectric effect in potting resins", 12th Electrical/Electronics Insulation Conference, 1975, p. 52–7, EEA79–36690.

2696. Jakobsen, R. J., "Detection and characterization of water-induced reversion of epoxy and urethane potting compounds", Report AD–A057316/2SL, Battelle Columbus Labs., Ohio, 1978, 77 pp.

2697. Kahnau, H. W. and Kieninger, W., "Fire-safety of thermoplastic mouldings", Elektrotech. Z. (ETZ) B, 28 (1), 2–9 (1976), German, EEA79–11897.

2698. Kakei, M. and Ikeda, Y., "Transfer molding methods expand packaging techniques", JEE, no. 126, 59–62 (1977), EEA81–27205.

2699. Kale, V. S., "Interaction of parylene and moisture in hermetically sealed hybrids", Proceedings of the 28th Electronic Components Conference, 1978, p. 344–9, EEA81–42162.

2700. Kao, K. C., "Theory of high-field electric conduction and breakdown in dielectric liquids", IEEE Trans. Electr. Insul., EI–11 (4), 121–8 (1976), EEA80–3564.

2701. Kape, J. M., "Testing sealing quality of anodic coatings. I.", Finish. Ind., 1 (11), 13, 16–18 (1977), EEA81–4784.

2702. Kern, W. and Comizzoli, R. B., "New methods for detecting structural defects in glass passivation films", J. Vac. Sci. Technol., 14 (1), 32–9 (1977).

2703. Kern, W. and Comizzoli, R. B., "Techniques for measuring the integrity of passivation overcoats on integrated circuits", Natl. Bur. Stand. Spec. Publ., no. 400–31, 105 pp. (1977), CA86–198778.

2704. King, G. R., "Quality control and screening in the production of plastic encapsulated semiconductor devices (PEDs)", Microelectron. & Reliab., 16 (3), 245-9 (1977), EEA81-135.

2705. Koppe, R., "Influence of defects on the electrical ageing of epoxy resin insulations", Elektrie, 30 (1), 19-20 (1976), German, EEA79-18764.

2706. Kozakiewicz, R. P. and Sbar, N. L., "Procedures for comparative evaluation of the moisture protection of coatings for integrated circuits", Electrochemical Society Spring Meeting, Extended Abstracts, 1977, p. 123-4, EEA81-5612.

2707. Krammy, W., "The leak testing of components - represented by the example of hermetically sealed relays", Mess. & Pruef., no. 11, 652-6 (1976), German, EEA80-8580.

2708. Lamson, M. A. and Ramsey, T. H., Jr., "Improving the quality of glass-sealed ceramic DIPs", Electron. Packag. & Prod., 18 (1), 84-96 (1978).

2709. Levinthal, D. J., "Leak detection and the matter of interpretation", Electron. Packag. & Prod., 17 (5), 37-8, 40-4 (1977).

2710. Littlewood, S. and Briggs, B. F. N., "Investigation of current-interruption by metal-filled epoxy resin", J. Phys. D, 11 (10), 1457-62 (1978), EEA81-42113.

2711. Lovell, R., "The effect of specimen size of the electric breakdown of unfilled and filled epoxy polymers", IEEE Trans. Electr. Insul., EI-11 (4), 110-4 (1976), EEA80-3563.

2712. McFarland, J. W., "Chemical and physical characterization of a Kelvar-filled epoxy molding compound", Report BDX-613-1816, Bendix Corp., Kansas City, Mo., 1978, 16 pp.

2713. McSheehy, W. H., "Additive enhances hydrolytic stability of polyester-based urethane rubber potting and molding compound", Insul./Circuits, 22 (10), 33-6 (1976), EEA80-43.

2714. Mara, J. F. and Sergent, J. E., "Electrical properties of epoxies used in hybrid microelectronics at low microwave frequencies", Proceedings of the 1977 International Microelectronics Symposium, p. 170-4.

2715. Mark, R. and Findley, W. N., "Thermal expansion instability and creep in amine-cured epoxy resins", Polym. Eng. Sci., 18 (1), 6-15 (1978).

2716. Martin, W. L., Jr., "Class H solventless encapsulants - unsaturated polyester development and evaluation", Proceedings of the 13th Electrical/Electronics Insulation Conference, 1977, p. 309-12, EEA81-13308.

2717. Matsuoka, S., Sunaga, H., Tanaka, R., Hagiwara, M. and Araki, K., "Accumulated charge profile in polyethylene during fast electron irradiations", IEEE Trans. Nucl. Sci., NS-23 (5), 1447-52 (1976), EEA80-3562.

2718. Nagai, Y., Suzuki, T. and Wakabayashi, Y., "Development and evaluation of epoxy resin molding compounds for encapsulation use", Fujitsu Sci. Tech. J, 11 (2), 115-33 (1975).

2719. Noskov, A. M. and Novikov, N. I., "Quantitative spectroscopic description of the photodegradation of hardened copy oligomers", J. Appl. Spectrosc., 26 (6), 768–72 (1977).

2720. Okada, K. and Tanahashi, M., "Plastic package seal test by $^{85}$Kr radioactive tracer techniques", Natl. Tech. Rep., 23 (1), 184–90 (1977), Japanese, EEA80-40324.

2721. Planting, P. J., "An approach for evaluating epoxy adhesives for use in hybrid microelectronic assembly", IEEE Trans. Parts, Hybrids & Packag., PHP-11 (4), 305–11 (1975), EEA79-13052.

2722. Potapov, A. I., Rapoport, D. A. and Klopov, V. D., "A mechanized unit for infrared inspection of plastic parts", Sov. J. Nondestr. Test., 13 (2), 212–15 (1977), EEA81-41424.

2723. Pugacz-Muraszkiewicz, I. J. and Hammond, B. R., "Application of silicates to the detection of flaws in glassy passivation films deposited on silicon substrates", J. Vac. Sci. Technol., 14 (1), 49–53 (1976).

2724. Quach, A. and Hunter, W. L., "Study of properties of plastics used for semiconductor encapsulation", J. Electron. Mater., 6 (3), 319–31 (1977), EEA81-19396.

2725. Ramamurti, T. V., "Role of mica in electronics industry", Electron. Inf. & Plann., 3 (10), 761–70 (1976), EEA80-687.

2726. Roberts, B. C., "Evaluation of an MOD specification for the procurement of plastic encapsulated semiconductor devices", Proceedings of the Symposium on Plastic Encapsulated Semiconductor Devices, 1976, p. 6/1–7, EEA80-6606.

2727. Robinson, D. D., "Evaluation testing of solid encapsulated microcircuits for space applications", Report N77-25556/0SL, Boeing Aerospace Co., Seattle, Wash., 1977, 179 pp.

2728. Ruthberg, S., Neff, G. R. and Martin, B. D., "Radioisotope hermetic test precision", Proceedings of the 1977 International Microelectronics Symposium, p. 131–7.

2729. Schroter, S., "Tests and evaluation of polymers for hybrid assembly", 12th Electrical/Electronics Insulation Conference, 1975, p. 68, EEA79-37740.

2730. Sergent, J. E., Stout, C. W. and Caruso, S. V., "Electrical properties of adhesives used in hybrid microelectronic applications", 26th Electronic Components Conference, 1976, p. 373–5, EEA79-37731.

2731. Sundgvist, B., Sandberg, O. and Backstrom, G., "Thermal properties of an epoxy resin at high pressure and temperature", J. Phys. D, 10 (10), 1397–1403 (1977).

2732. Szedon, J. R., "Tests for electrically active surface contaminants from hybrid microcircuit adhesives", 26th Electronic Components Conference, 1976, p. 368–72, EEA79-37730.

2733. Taylor, J., "High voltage insulation testing", Electron, no. 94, 62 (1976), EEA79-34932.

2734. Teranishi, T., Ohashi, A. and Ueda, M., "Dielectric characteristics of silicone oils at low temperature", Electr. Eng. Jap., 95 (2), 1–5 (1975), EEA79-16912.

2735. Theocaris, P. S., Paipetis, S. A. and Papanicolaou, G. C.,
      "Indentation studies in plasticized epoxy polymers", J. Appl.
      Polym. Sci., 22 (5), 1417-30 (1978).
2736. Theocaris, P. S., Paipetis, S. A. and Papanicolaou, G. C.,
      "Indentation studies in aluminum-filled epoxies", J. Appl.
      Polym. Sci., 22 (8), 2245-52 (1978).
2737. Theocaris, P. S. and Prassinakis, J., "Interrelations of
      mechanical and optical properties of plasticized epoxy poly-
      mers", J. Appl. Polym. Sci., 22 (6), 1725-34 (1978).
2738. Thies, C., "Physicochemical aspects of microencapsulation",
      Polym. Plast. Technol. Eng., 5, 1-22 (1976).
2739. Thomas, J., "Procedure for humidity testing of plastic encap-
      sulated ICs", Radio Fernsehen Elektron., 27 (6), 388-90
      (1978), German, EEA82-117.
2740. Traeger, R. K., "Nonhermeticity of polymeric lid sealants",
      IEEE Trans. Parts, Hybrids & Packag., PHP-13 (2), 147-52
      (1977), EEA80-31396.
2741. Vanderkooi, N. and Riddell, M. N., "Dynamic permeability
      method for epoxy encapsulation resins", 14th Annual Proceed-
      ings Reliability Physics Symposium, 1976, p. 219-22.
2742. Venning, B. H., "High frequency performance of viscoelastic
      packaging materials", J. Soc. Environ. Eng., 15-2 (69), 26-9
      (1976), EEA79-31896.
2743. Wales, R. D. and Kostinko, W., "Microhardness measurements on
      as-plated coatings", Plat. & Surf. Finish., 64 (3), 30-4
      (1977), EEA81-9266.
2744. Walker, J. M., Richardson, W. E. and Smith, C. H., "CTBN-
      modified epoxy encapsulant", Mod. Plast., 53 (5), 62-4 (1976).
2745. Wille, G., "Property comparison of flexible packaging films",
      Plaste Kautsch., 23, 670-2 (1976), German, CA85-178543.
2746. Woodard, J. B., "Thermally stimulated discharge in fractional
      component and moisture effect studies of epoxy encapsulants",
      J. Electron. Mater., 6 (2), 145-62 (1977), EEA81-133.
2747. Yang, H. W. and Tyler, A. L., "The formation of current leak-
      age paths by diffusion of water through protective coatings",
      Ind. & Eng. Chem., Prod. Res. & Dev., 16 (3), 252-7 (1977),
      EEA81-5615.
2748. Young, P. L., Fehlner, F. P. and Whitman, A. J., "Correlation
      of sputtering conditions with electronic conduction in $Ta_2O_3$
      films", J. Vac. Sci. & Technol., 14 (1), 174-6 (1977),
      EEA80-20578.
2749. Yurechko, N. A., Lipskaya, V. A., Shologon, I. M., et al,
      "Effect of the position of functional groups in the aromatic
      nucleus on the mechanical properties of hardened epoxy resins",
      Polym. Sci., 19 (2), 413-19 (1977).
2750. Zucconi, T. D., "Liquid chromatography detection for epoxy
      resins", IBM Tech. Disclosure Bull., 20 (12), 5140 (1978).
2751. Zurilla, R. W. and Hospadaruk, V., "Quantitative test for zinc
      phosphate coating quality", SAE Prepr. no. 780187, 1978, 8 pp.

## 5.  Integrated Circuits

2752. Barber, M. R. and Zacharias, A., "Integrated circuit testing",
      Bell Lab. Rec., 124-30 (1977), EEA80-32450.
2753. Barroeta, J. J., Camps, S., Suarez, R. E., Canas, M. A. and
      Amaya, R. A., "Fault-detection system for digital integrated
      circuits", IEEE Trans. Instrum. Meas., IM-26 (3), 246-51 (1977).
2754. Beniere, F., "New method of microanalysis of electronic com-
      ponents", Rev. Phys. Appl., 12 (11), 1805-9 (1977), French,
      EEA81-19522.
2755. Bisset, S., "Distributed microcomputers in LSI test system net-
      works", Digest of Papers 1978 Semiconductor Test Conference,
      p. 196-9.
2756. Black, S., "Optimizing manual circuit-pack tests", Digital
      Des., 6 (10), 66, 68 (1976), EEA80-27964.
2757. Campbell, J. F., Jr., "A new real-time function test genera-
      tion system for complex LSI testing", 1975 WESCON Technical
      Papers, Vol. 19, Paper 27/2, 4 pp., EEA79-15931.
2758. Campbell, J., Johnson, E. and Taylor, A. R., "A new software
      system for LSI testing", Digest of Papers 1977 Semiconductor
      Test Symposium, p. 131-4, EEA81-19528.
2759. Chi, C. S., "Intelligent strategy with distributed database
      for high-volume LSI testing", IEEE Trans. Instrum. Meas.,
      IM-27 (2), 172-8 (1978).
2760. Child, M. R., Ramaspaghe, D. W. and White, D. W., "Dynamic
      inspection of large scale integrated circuits", Microelec-
      tronics, 7 (3), 43-8 (1976), EEA79-31849.
2761. Chrones, C., "Calculating the cost of testing LSI chips",
      Electronics, 51 (1), 171-3 (1978).
2762. Dance, M., "Evaluating passive and hybrid IC's", Electron.
      Ind., 3 (12), 27-31, 33, 35, 37 (1977), EEA81-10046.
2763. Edwards, W. D., Smith, J. G. and Kemhadjian, H. A., "Some
      investigations into optical probe testing of integrated cir-
      cuits", Radio & Electron. Eng., 46 (1), 35-41 (1976),
      EEA79-16826.
2764. Esposito, R. M., Garbarino, P. L., Hallas, N. E., Levine, J.
      E., O'Reilly, F. J. and Satya, A. V., "Test site for a semi-
      conductor masterslice", IBM Tech. Disclosure Bull., 20 (4),
      1418-19 (1977), EEA81-24037.
2765. Goldstein, L., "Computational complexity/confidence level
      trade-offs in LSI testing", Digest of Papers 1978 Semicon-
      ductor Test Conference, p. 50-8.
2766. Golik, S. M., Perepelitsyn, N. L. and Dorodko, A. F., "Machine
      method of determining monitoring tests for large integrated
      circuits", Mekh. & Avtom. Upr., no. 2, 49-52 (1976), Russian,
      EEA79-33085.
2767. Gopinath, A. and Gopinathan, K. G., "High-resolution sampling
      SEM for quantitative investigations of semiconductor devices
      and integrated circuits", IEEE Trans. Electron. Devices,
      ED-25 (4), 431-4 (1978).

2768. Grimm, A. V., "LSI integrated circuit test development",
      Report BDX-613-1017, Bendix Corp., Kansas City, Mo., 1974,
      62 pp., EEA78-24509.
2769. Hall, C. M., Jr., Shade, E. E. and Shukis, J. R., "A compara-
      tive evaluation of IC packages in commercial real-time computer
      terminals", Proceedings 1976 Annual Reliability and Maintain-
      ability Symposium, p. 170-5, EEA79-46362.
2770. Hayne, G. S. and Beadles, R. L., "Automation feasibility study
      for microelectronic wafer and integrated circuit test set,
      TTU-311/E", Report AD-A017322/9GA, Res. Triangle Inst. Res.,
      Triangle Park, N. C., 1975, 93 pp., EEA79-24893.
2771. Healy, J., "Is it really necessary to test LSI?", Electron,
      no. 118, 18, 20, 23 (1977), EEA81-9255.
2772. Healy, J., "An analysis of trends in complex LSI testing stra-
      tegies", Digest of Papers 1978 Semiconductor Test Conference,
      p. 59-64.
2773. Hnatek, E. R., "Test results from screening linear integrated
      circuits", Eval. Eng., 14 (5), 32-3 (1975), EEA79-3767.
2774. Hnatek, E. R., "User's tests, not data sheets, assure IC per-
      formance (digital LSI)", Electronics, 48 (24), 108-13 (1975),
      EEA79-4747.
2775. Homan, R. A. and Rossman, M. W., "Evaluation testing of inte-
      grated circuits", Proceedings of the 1975 Annual Reliability
      and Maintainability Symposium, p. 372-6, EEA79-46393.
2776. Jenkins, C. R. and Durgin, D. L., "EMP susceptibility of inte-
      grated circuits", IEEE Trans. Nucl. Sci., NS-22 (6), 2494-9
      (1975).
2777. Klein, R. D., "Simultaneous display of all digital IC levels
      on a single oscilloscope", Elektronik, 25 (10), 88-9 (1976),
      German, EEA80-670.
2778. Knowlton, D., "Is there a future for distributed systems?
      [LSI ATE]", Digest of Papers 1977 Semiconductor Test Symposium,
      p. 136-7, EEA81-19529.
2779. Kolodziejski, J., "Quick testing of integrated circuits",
      Wiad. Telekomun., 18 (1), 4-8 (1978), Polish, EEA81-33013.
2780. Kottek, E. and Tomanova, B., "IC tests of the MH74 and MH74S
      series based on structure", Sdelovaci Tech., 24 (5), 163-4
      (1976), Czech, EEA79-41600.
2781. Kreinberg, W., "Testing large scale integrated circuits of
      own design supplied by another manufacturer", Elektron. Anz.,
      8 (3), 71-3 (1976), German, EEA79-28663.
2782. Laliotis, T. A., "Application of microcomputers to low cost
      digital IC testers", International Microelectronics Conference,
      Proceedings of the Technical Program, 1975, p. 204-12.
2783. McDonald, J. C. and McCracken, P. T., "Testing for high reli-
      ability (digital ICs)", 14th IEEE Computer Society Interna-
      tional Conference, 1977, p. 190-1, EEA81-1308.

2784. Madsen, D., "Controlled generation and manipulation of asyn-
      chronous inputs to LSI devices in low cost general purpose LSI
      test systems", Digest of Papers 1978 Semiconductor Test Con-
      ference, p. 186-91.
2785. Magrisso, I. B. and Sharp, L. A., "Electrical partitioning
      technique to test LSI packages", IBM Tech. Disclosure Bull.,
      21 (1), 7-9 (1978).
2786. Nelson, G. F. and Boggs, W. F., "Parametric tests meet the
      challenge of high-density ICs", Electronics, 48 (25), 108-11
      (1975), EEA79-8998.
2787. Obright, H. L., "Electrical testing of bare silicon chips prior
      to hybrid fabrication", 27th Electronic Components Conference,
      1977, p. 141-5, EEA80-35946.
2788. Ohtani, Y., "IC test system - a look at the state of the art",
      JEE, no. 110, 12-15 (1976), EEA79-19610.
2789. Petrenko, A. I., Tsurin, O. F., Savatev, V. A., Bobovskii, V.
      V. and Tetelbaum, A. Y., "Using a graphic design for inspect-
      ing the topology of large integrated circuits", Mekh. & Avtom.
      Upr., no. 6, 46-9 (1975), Russian, EEA79-20541.
2790. Purkiss, N., "Testing LSI", New Electron., 10 (22), 109 (1977),
      EEA81-19521.
2791. Reitan, A. and Silkoset, O., "Test and acceptance of LSI cir-
      cuits", Elektro, 90 (22), 8, 10 (1977), Norwegian, EEA81-19512.
2792. Rezvyi, R. R., Buiko, L. D., Kontsevoi, Y. A. and Pilipenko,
      V. A., "The use of ellipsometric monitoring methods at various
      stages of integrated circuit preparation processes", Sov.
      Microelectron., 5 (4), 268-71 (1976), EEA80-35901.
2793. Rice, J. C., "Base your IC tester on a µP", Electron. Des.,
      24 (1), 88-92 (1976).
2794. Runyon, S., "How to test those LSI chips: Watch out for 'ifs'
      and 'buts'", Electron. Des., 23 (24), 74-8 (1975), EEA79-13015.
2795. Schafft, H. A., "Semiconductor measurement technology: ARPA/
      NBS workshop-III. Test patterns for integrated circuits",
      Natl. Bur. Stand. Spec. Publ. No. 400-15, 1976, 52 pp.
2796. Schmid, J. H., "Semi-automatic function tester for integrated
      TTL circuits", Funkschau, 47 (19), 67-8 (1975), German,
      EEA79-1046.
2797. Schmitt, R., "Testing LSI analog chips", Digest of Papers 1978
      Semiconductor Test Conference, p. 18-22.
2798. Shulman, S. E., "Current tests ensure IC-package orientation",
      Electronics, 48 (25), 117, 19 (1975), EEA79-8999.
2799. Sigsby, B., "Incoming test the money saver", Electron. Equip.
      News, 12-13 (1975), EEA79-9001.
2800. Simon, G., "Examination of thermo-electrical interaction in
      analogue integrated circuits", Hiradastechnika, 27 (2), 33-6
      (1976), Hungarian, EEA79-24905.
2801. Smith, D., "A sane, self-educating approach to LSI test devel-
      opment", Digest of Papers 1978 Semiconductor Test Conference,
      p. 42-5.

2802. Smith, P. L. and Brown, W. C., "A simple minded software approach to IC testing", Digest of Papers 1977 Semiconductor Test Symposium, p. 119-24, EEA81-19526.

2803. Stetzler, G. F., Filek, M. and Youtz, R. E., "High-voltage production testing of linear integrated circuits", Western Electric Eng., 20 (1), 2-9 (1976), EEA79-16862.

2804. Stetzler, G. F., Filek, R. M. and Youtz, R. E., "Pin-oriented program, cassette controlled probe sustain high throughput and versatility in IC production test system", Insul./Circuits, 22 (6), 55-9 (1976), EEA79-37689.

2805. Yu, E. Y., "Determination of temperature of integrated-circuit chips in hybrid packaging", IEEE Trans. Parts, Hybrids & Packag., PHP-12 (2), 139-46 (1976).

## 6.   Substrates

2806. Barney, W. H., "The measurement and analysis of the dielectric strength of glasses", 1975 Dielectric Materials, Measurements and Applications, p. 281-4, EEA79-20577.

2807. Bentley, R. E. and Jones, T. P., "The isolation of bulk and interface components in the electric resistance of alumina", J. Aust. Ceram. Soc., 11 (2), 30-6 (1975), EEA79-24684.

2808. Borase, V. and Cambey, L. A., "A new 96 percent improved alumina substrate", 27th Electronic Components Conference, 1977, p. 400-3, EEA80-35926.

2809. Comeforo, J. E., "Selecting ceramic substrates for semiconductor devices", Insul./Circuits, 22 (13), 15-19 (1976), EEA80-13278.

2810. Deutsch, J. and Schlagheck, E., "A time saving method for inspection and measurement of the permitivity of ceramic substrates", Nachrichtentech. Z. (NTZ), 29 (2), 148-9 (1976), German, EEA79-16859.

2811. Dougherty, W. E., "Control of thermal coefficient of expansion of substrate materials", IBM Tech. Disclosure Bull., 19 (8), 3048 (1977), EEA80-41118.

2812. Esser, K. R., "Ultrasonic material testing - reproducible determination of material faults", Elektr. Ausruestung, 18 (4), 18-21 (1977), German, EEA81-396.

2813. Flinn, D. R., McDonough, W., Stern, K. H. and Rice, R., "Ion conductivity of hot-pressed sodium beta-alumina", Electrochemical Society Spring Meeting, Extended Abstracts, 1975, p. 23-4, EEA79-746.

2814. Fritz, L. L., "Physical defects in thick-film hybrid substrates", 26th Electronic Components Conference, 1976, p. 32-54, EEA79-37720.

2815. Fritz, L. L., "Ceramic evaluation and screening for a hi-rel thick-film large area hybrid application using a stringent dye penetrant method", Proceedings of the 1977 International Microelectronics Symposium, p. 228-34.

2816. Gerhard, A. R., "Measuring dielectric constant of substrates
      for microstrip applications", IEEE Trans. Microwave Theory
      Tech., MTT-24 (7), 485-7 (1976).
2817. Howard, R. T., "Semiconductor device package having a substrate
      with a coefficient of expansion matching/silicon", IBM Tech.
      Disclosure Bull., 20 (7), 2849-50 (1977), EEA81-45184.
2818. Igisu, T., "Method for removing the strain of a substrate for
      an element of integrated circuit", Patent USA 4026010, Publ.
      May 1977.
2819. Ikeda, H., Inagaki, T. and Nishimura, Y., "Ceramic surface in-
      spection using laser technique (hybrid IC substrates)", Jap.
      J. Appl. Phys., 14 (Suppl. 14-1), 487-92 (1975), EEA79-4782.
2820. Jentzsch, J., Ostwald, R. and Bogenschutz, A. F., "Investiga-
      tions on the thermal expansion of thick-film materials", Wiss.
      Ber. AEG-Telefunken, 49 (6), 229-34 (1976), EEA80-24619.
2821. Kitaev, M. A., "Measurement of the complex impedances of SHF
      microcircuits", Sov. Microelectron., 6 (1), 60-2 (1977),
      EEA82-648.
2822. Koga, K. and Enomoto, Y., "Fine-grained BaTiO3 for thick film
      dielectrics", Electrochemical Society Fall Meeting, Extended
      Abstracts, 1975, p. 361-2, EEA79-16869.
2823. Kurtz, J. and Semitschew, P., "Selecting microelectronic sub-
      strates for thick film and hybrid circuits", Circuits Manuf.,
      17 (10), 50-3 (1977).
2824. Lakubowski, W., "Electrical properties of aluminas", Elek-
      tronika, 17 (7-8), 261-4 (1976), Polish, EEA80-686.
2825. Libove, C., "Impact stresses in flat-pack lids and bases",
      Report AD-A054947/7SL, Syracuse Univ., N. Y., Dept. of
      Mechanical and Aerospace Engineering, 1978, 45 pp.
2826. Licznerski, B., Nitsch, K. and Rzasa, B., "Low frequency char-
      acteristics of TiO2 (rutile)-glass thick films", Electrocom-
      pon. Sci. & Technol., 4 (1), 1-7 (1977), EEA81-5619.
2827. Musgrave, W. J., "Evaluation of ceramic dip carrier", Report
      AD-A056653/9SL, Naval Avionics Center, Indianapolis, In.,
      1978, 85 pp.
2828. Niwa, K. and Murakawa, K., "New alumina substrates for micro-
      electronics", Electrocompon. Sci. & Technol., 2 (2), 115-19
      (1975), EEA79-4723.
2829. Niwa, K., Yokoyama, H. and Murakawa, K., "Mechanical strength
      of fine grained alumina (FGA) substrate", Fujitsu Sci. &
      Tech. J., 13 (1), 87-100 (1977), EEA80-32234.
2830. Podhoranyi, I. and Sztankovics, L., "Alumina ceramics for
      electronic and electrical purposes", Hiradstech. Ipari Kut.
      Intez. Kozl., 15 (3), 34-45 (1975), Hungarian, EEA79-20574.
2831. Przybysz, E., "Features of some substrate and conducting
      paths materials in 250-700 MHz frequency band", Elektronika,
      18 (4), 154-6 (1977), Polish, EEA80-28627.
2832. Rose, A., Ugol, L. and Gelb, A., "Resistor compatible multi-
      layer dielectrics", 12th Electrical/Electronics Insulation
      Conference, 1975, p. 70-4, EEA79-37652.

2833. Roth, W. L., "Crystal chemical studies of sodium ion conduc-
      tion in beta-alumina", Electrochemical Society Spring Meeting,
      Extended Abstracts, 1975, p. 17-18, EEA79-745.
2834. Titeux, M., "A solution of the future: the sapphire substrate",
      Toute Electron., no. 416, 35-8 (1976), French, EEA80-9204.
2835. Traut, G. R., "Electrical measurements of microwave circuit
      board materials", Proceedings of the 13th Electrical/Elec-
      tronics Insulation Conference, 1977, p. 39-44, EEA81-14246.
2836. Walsh, K. A., "Manufacture of super fine finish beryllia",
      27th Electronic Components Conference, 1977, p. 404-7,
      EEA80-35902.
2837. Westphal, K., "Substrate materials for the integrated micro-
      wave technique", Nachrichtentech. Elektron., 26 (11), 403-7
      (1976), German, EEA80-3307.
2838. Yamada, S. and Murakawa, K., "High density alumina ceramics
      for microelectronics devices", Patent Japan 76-10813, Publ.
      January 1976, CA85-27874.
2839. Zurbrick, J. R., "Techniques for nondestructively character-
      izing metallic substrate surfaces prior to adhesive bonding",
      In: Int. Adv. Nondestr. Test. Vol. 5, Gordon & Breach, New
      York, p. 41-70 (1977).

## 7.  Others

2840. Alessandri, A., "Quality control and testing of electronic
      components and sub-assemblies", Autom. and Strum., 26 (1),
      18-20 (1978), Italian, EEA81-27206.
2841. Alius, E., "Selection of (lead frame) metallic comb- and
      supporting strip-materials respectively for DIL-ceramic and
      plastic packages", Feingeraete Tech., 25 (10), 452-5 (1976),
      German, EEA80-9497.
2842. Allington, T. R. and Cote, R. E., "Characterization of thick
      film compositions on porcelain steel substrates", Solid State
      Technol., 22 (1), 81-6 (1979).
2843. Anon., "Micro-technology uses fluorochemical for simple, reli-
      able stress tests", Eval. Eng., 16 (3), 56-7 (1977),
      EEA80-35913.
2844. Balenovich, J. D. and Karstaedt, W. H., "Induction heating
      effects on semiconductor disc packages", 10th IAS (IEEE Ind.
      Appl. Soc.) Annual Meeting, Conference Record, 1975, p. 340-2.
2845. Band, K. S., Sherwood, B. J., Scobey, I. H. and Ward, R. C.
      C., "Characterization of metal films on corundum substrates
      (MIC applications)", J. Mater. Sci., 12 (3), 577-82 (1977),
      EEA80-20584.
2846. Bartels, A. L., "The relationship between stress and strain
      in flameproof enclosures", 2nd International Conference on
      Electrical Safety in Hazardous Environments, 1975, p. 153-7,
      EEA79-24049.

2847. Beaven, P. A. and Brown, D. J., "Measuring the impedance of a three-state bus", IBM Tech. Disclosure Bull., 21 (5), 2016 (1978).

2848. Breton, P. J. and Archuletta, M. D., "SEM inspection in electronics", Circuits Manuf., 18 (12), 44-7 (1978).

2849. Bullis, W., "Semiconductor measurement technology", Report AD-A040011/9SL, National Bureau of Standards, Washington, D. C., 1976, 122 pp.

2850. Burns, D. J., "Microcircuit analysis techniques using field-effect liquid crystals", IEEE Trans. Electron. Devices, ED-26 (1), 90-5 (1979).

2851. Czechowski, A., "Comparison of methods for testing dynamic strength of electronic components and equipment", Pomiary Autom. Kontrola, 23 (6), 218-21 (1977), Polish, EEA81-13282.

2852. David, R. F. S., "Practical limitations of PIND testing", Proceedings of the 28th Electronic Components Conference, 1978, p. 281-5, EEA81-37930.

2853. Dobrott, R. D., Keenan, J. A. and Larrabee, G. B., "Ion micro-probe analysis of integrated circuit structures", 15th Annual Proceedings Reliability Physics Symposium, 1977, p. 54-60, EEA80-41551.

2854. Dodge, H. F., "Inspection for quality assurance", J. Qual. Technol., 9 (3), 99-101 (1977), EEA81-4787.

2855. Dreyer, H., "Drawing up optimum testing program in electronic technology", Nachrichtentech. Elektron., 28 (4), 151-4 (1978), German.

2856. Droscha, H., "Rationalised quality control by fully automatic laser-operated surface inspection", Kunstst.-Berat., 21 (11), 626-7 (1976), German, EEA80-16457.

2857. Dye, R. and Woodmancy, G. C., "On-line inspection system assures 100% foreign particle free packages", Package Dev. Syst., 7 (1), 17-19 (1977).

2858. Dyer, D. C., Swanson, G. D. and Walters, D. E., "Technique completely quantifies stress in encapsulated component lead wires", Insul./Circuits, 23 (9), 38-40 (1977).

2859. Franco, J. V., Mariano, M. H. and Rave, W. C., "Capacitor/thin-film stress test circuit", IBM Tech. Disclosure Bull., 18 (3), 813-14 (1975), EEA79-1029.

2860. Guertin, R., "Hermeticity test pinpoints leaks in thick film hybrids without damaging components", Insul./Circuits, 23 (2), 26 (1977).

2861. Guidici, D. C., "Stereo radiography of complex multilayered microcircuits", Solid State Technol., 19 (10), 45-6, 49 (1976), EEA80-3308.

2862. Hathaway, J. W. and Yu, C. C., "Thermal fatigue testing of discrete components", IBM Tech. Disclosure Bull., 19 (11), 4269-70 (1977), EEA81-13276.

2863. Hellstrom, S. and Wesemeyer, H., "Nonlinearity measurements of thin films", Vacuum, 27 (4), 339-43 (1977), EEA81-5617.

2864. Himmel, R. P. and Kamensky, A., "Characteristics of thick-film conductors and resistors on beryllia substrates", Proceedings of the 1977 International Microelectronics Symposium, p. 8-15.

2865. Hotchkiss, J., "The impact of in-circuit prescreening on functional board testing", Circuits Manuf., 17 (10), 54, 58, 60-6 (1977).

2866. Hu, S. M., "Cathodic mapping of leakage defects (ICs)", J. Electrochem. Soc., 124 (4), 578-82 (1977), EEA80-24901.

2867. Koga, Y. and Sasaki, I., "Methods of wiring check and evaluation of their validity", Inf. Process. Soc. Jap. (Joho Shori), 17 (8), 711-19 (1976), Japanese, EEA80-9023.

2868. Lawson, R. W., "The qualification approval of plastic encapsulated semiconductor components for use in moist environments", Proceedings of the Symposium on Plastic Encapsulated Semiconductor Devices, 1976, p. 7/1-6, EEA80-6607.

2869. Libove, C., "Rectangular flat-pack lids under external pressure", Report AD-A025625/5SL, Syracuse Univ., N. Y., 1976, 68 pp.

2870. Lin, K. and Burden, J. D., "Mass spectrometric solutions to manufacturing problems in the semiconductor industry", J. Vac. Sci. & Technol., 15 (2), 373-6 (1978), EEA81-41419.

2871. Lord, H. and Barbero, L. P., "Problems arising from the use of organic plastics materials in flameproof enclosures", 2nd International Conference on Electrical Safety in Hazardous Environments, 1975, p. 59-64, EEA79-24048.

2872. McCullough, R. E., "Hermeticity and particle impact noise test techniques", 14th Annual Proceedings Reliability Physics Symposium, 1976, p. 256-62, EEA80-16448.

2873. Manka, W. V., "Alternative methods for determining chip inductor parameters", IEEE Trans. Parts, Hybrids & Packag., PHP-13 (4), 378-85 (1977), EEA81-14193.

2874. Mann, J. E., et al., "Test proposals for improved electronics production", Circuits Manuf., 17 (7), 44-53 (1977).

2875. Mari, C., Scolari, V., Fiori, G. and Pizzini, S., "Structural electrical and electrochemical characterization of Ni-Pr oxide thick films", J. Appl. Electrochem., 7 (2), 95-106 (1977).

2876. Mentley, D. E., "Microstructure of a base metal thick film system", Report LBL-5176, California Univ., Berkeley, 1976, 36 pp.

2877. Morabito, J. M., Thomas, J. H., III and Lesh, N. G., "Material characterization of Ti-Cu-Ni-Au (TCNA) - a new low cost thin film conductor system", IEEE Trans. Parts, Hybrids & Packag., PHP-11 (4), 253-62 (1975).

2878. Nishida, N., Suzuki, H., Kosuge, K., Koreeda, S. and Fukui, S., "Holographic observation of deformation in header used in plastic package for high output power transistor", NEC Res. & Dev., no. 46, 33-41 (1977), EEA81-4778.

2879. O'Dell, G. D., "Necessity of characterizing hybrid microcir-
      cuit elements at microwave frequencies", Report BDX-613-1477,
      Bendix Corp., Kansas City, Mo., 1976, 13 pp.
2880. Olevskii, S. S., Gorokhov, V. N. and Aleksandrov, G. V., "Use
      of Auger spectroscopy for chemical analysis of thin film
      structures", Prib. & Sist. Upr., no. 6, 46-7 (1978), Russian,
      EEA81-49500.
2881. Peter, A. E., "Method for reducing connector crosstalk", IBM
      Tech. Disclosure Bull., 21 (3), 958 (1978).
2882. Piccoli, S. C., "Strength tests of 1-mil diameter aluminum
      wire exposed to high temperature for various time periods",
      Proceedings of the 1977 International Microelectronics Sym-
      posium, p. 150-4.
2883. Remshardt, H., Schettler, H., Schumacher, H. and Zuehlke, R.,
      "Tolerance trimming of electrical data large scale integra-
      tion semiconductor circuits", IBM Tech. Disclosure Bull., 20
      (8), 3189-90 (1978).
2884. Sander, W., "Testing of components with perfluorinated
      liquids", Elektronik, 26 (4), 93-4 (1977), German, EEA80-31391.
2885. Schaefer, G., Ward, R. and Zakraysek, L., "Evaluation of lead
      finishes for microcircuit packages", Report AD-A015723/0SL,
      General Electric Co., Utica, N. Y., 1975, 140 pp.
2886. Schneider, B., Zloof, H., Vinikman, V. and Samuel, A., "More
      about the microstructure of thick film materials", 9th Con-
      vention of Electrical and Electronic Engineers in Israel,
      1975, p. C2-2/1-13, EEA79-46400.
2887. Slesarev, A. I., Kortov, V. S., Kordyukov, N. I. and Plekah-
      nova, E. A., "Use of exoemission to monitor the quality of
      microcircuit elements", Sov. Microelectron., 5 (6), 433-5
      (1977), EEA81-906.
2888. Stevens, R. E. and Martens, A. E., "The happy marriage of the
      microscope and electronics", Automation, 23 (12), 42-5 (1976),
      EEA80-19692.
2889. Thomas, R. W., "Moisture myths and microcircuits", 26th Elec-
      tronic Components Conference, 1976, p. 272-6, EEA79-37627.
2890. Thomas, R. W., "Microcircuit package gas analysis", 14th
      Annual Proceedings Reliability Physics Symposium, 1976,
      p. 283-94, EEA80-17033.
2891. Vainshtein, M. Z. and Plyshevskaya, E. M., "Effect of certain
      factors on the strength of thick oxide films formed during
      anodizing of aluminum alloys", Prot. Met., 12 (6), 639-41
      (1976).
2892. Weigand, B. L., Licari, J. J. and Pratt, I. H., "Detection of
      conductive condensates resulting from adhesive outgassing in
      hybrid microcircuits", Proceedings of the 28th Electronic
      Components Conference, 1978, p. 217-22, EEA81-37926.

XXIV. AUTOMATION

2893. Anderson, R. E., Fulks, R. G., Frusterio, C. P., Meade, F. S. and Phelps, D. E., "Processor-based tester goes on site to isolate board faults automatically", Electronics, 51 (10), 111-17 (1978).

2894. Anolick, E. S., Camenga, R. and Carlo, A., "Automatic ramper for simultaneous monitoring of many capacitors", IBM Tech. Disclosure Bull., 20 (11), 4491-2 (1978).

2895. Anon., "Thermal constraints in automatic soldering", Toute Electron., no. 416, 47-51 (1976), French, EEA80-9185.

2896. Anon., "Automatic wire bonding vs competing interconnection technologies", Electron. Prod., 7 (3), 53, 55 (1978), EEA81-49507.

2897. Anon., "Advances in automatic bonding", Circuits Manuf., 18 (10), 38, 40, 42 (1978).

2898. Anon., "Automatic testing [LSI circuit production]", Eng. Mater. & Des., 22 (6), 56-7 (1978), EEA81-41418.

2899. Atkinson, R. W., "Considerations for automation in thick-film screening", Circuits Manuf., 17 (4), 18, 20-2, 24, 26 (1977), EEA81-5618.

2900. Balekdjian, K. G., "Hardware-software tradeoffs for analog automatic test instruments", IEEE Trans. Instrum. Meas., IM-26 (3), 201-3 (1977).

2901. Belyakov, A. I., Kletskov, E. L., Panchenko, S. N. and Ovcharov, G. I., "Automatic monitoring of the ultrasonic welding process", Weld. Prod., 23 (7), 54-6 (1976), EEA80-24001.

2902. Bertails, J. C. and Zirphile, J., "A fast, reliable integrated circuit design method with automatic layout control", Rev. Tech. Thomson-CSF, 9 (4), 717-35 (1977), French, EEA81-23604.

2903. Bette, H. P., "Automatic testing of integrated circuits", Electron. Power, 23 (5), 380-4 (1977).

2904. Beyerlein, F. W., "New developments in automatic wire bonding equipment", Electron. Packag. & Prod., 18 (1), 54-6, 58-60, 64-7 (1978), EEA81-45767.

2905. Bjorklund, C. M. and Pavlidis, T., "On the automatic inspection and description of printed wiring boards", Proceedings of the International Conference on Cybernetics and Society, 1977, p. 690-3, EEA81-23576.

2906. Booth, R. H., Koppenhaver, B. L., Monahan, J. J. and Schneider, F. J., "Programmed multichip bonder", Tech. Dig., no. 46, 5-6 (1977), EEA81-14637.

2907. Bozoyan, S. E., "Construction of reliable circuits realizing a finite automaton from unreliable components", Eng. Cybern., 15 (2), 79-87 (1977).

2908. Burbank, K. E. and Reckhow, D. R., "Automatic test equipment translator board", IBM Tech. Disclosure Bull., 21 (4), 1404-5 (1978).

2909. Cochran, T. J. and Haas, R. G., "Automated punch apparatus
      for forming via holes in a ceramic green sheet", IBM Tech.
      Disclosure Bull., 20 (4), 1379-80 (1977), EEA81-22836.
2910. Crosher, R., "Is automatic inspection synonymous with auto-
      matic test?", Electron, no. 118, 29-30 (1977), EEA81-9256.
2911. Day, D. B., "Automatic test equipment test package develop-
      ment process", NAECON '78, Vol. 3, 1978, p. 1155-9.
2912. Devitt, D. and George, J., "Beam tape pins automated handling
      cuts IC manufacturing costs", Electronics, 51 (14), 116-19
      (1978), EEA81-42164.
2913. Fabenyi, E., "Automatic resistor trimming device for thick
      film circuits", Finommech.-Mikrotech., 17 (2), 59-63 (1978),
      Hungarian, EEA81-37913.
2914. Foley, E. B., Jr. and Firman, A. H., "Testing microcomputer
      boards automatically", Comput. Des., 15 (12), 92-4 (1976).
2915. Frodl, O., "Mechanisation of wire bonding for semiconductor
      devices", Feinwerktech. & Messtech., 85 (6), 243-5 (1977),
      German, EEA81-1225.
2916. Gaspar, J. Z., "Automatic measurement of integrated operational
      amplifiers on adaptor cards connected to automatic digital
      measuring equipment", Hiradstech. Ipari Kut. Intez. Kozl.,
      16 (1), 12-22 (1976), Hungarian, EEA79-37678.
2917. Gault, J. W. and Clary, J. B., "Application of microcomputers
      as on-line built-in-test elements", Proc. SOUTHEASTCON Reg.
      3 Conf., 1978, p. 462-5.
2918. Gllaher, J. B., "Development of an automated encapsulation
      system", Report BDX-613-1493, Bendix Corp., Kansas City, Mo.,
      1977, 29 pp.
2919. Griffiths, G. W., "Handling medium quantity production thick
      film hybrids", New Electron., 10 (6), 72, 74-5 (1977),
      EEA80-35934.
2920. Hanstead, P. D., "A new method for ultrasonic inspection",
      CEGB Tech. Disclosure Bull., no. 291, 1-5 (1978), EEA81-45190.
2921. House, D. and Steele, R. C., "Automated pin assembly system",
      IBM Tech. Disclosure Bull., 20 (7), 2616-18 (1977), EEA81-45770.
2922. Howard, J. S. and Nahourai, J., "Improvement in LSI produc-
      tion using an automated parametric test system", Solid State
      Technol., 21 (7), 48-52 (1978).
2923. Iddon, P., "Soldering methods for automatic production",
      Electron, no. 118, 33 (1977), EEA81-10058.
2924. Imamura, Y., Tanida, T. and Koga, M., "Application of micro-
      computer to fully-automatic die-bonder", Natl. Tech. Rep.,
      23 (6), 1102-11 (1977), Japanese, EEA81-28442.
2925. Khadpe, S., "Automated wire bonding verus tape automated
      bonding: what are the tradeoffs?", Proceedings of the 28th
      Electronic Components Conference, 1978, p. 402-8, EEA81-42173.
2926. Koontz, D. E. and Helgesen, G. F., "Continuous plating of
      ESS circuit pack receptacle connectors. I. A new approach to
      high-speed selective electroplating", Western Electric Eng.,
      22 (2), 26-31 (1978), EEA81-45707.

2927. Kowaichuk, R., Chen, Y. S. and Murray, J. L., "Continuous plating of ESS circuit pack receptacle connectors. III. Programmable control and status display for a dual strip-plating facility", Western Electric Eng., 22 (2), 41-7 (1978), EEA81-45709.

2928. Lang, J. and Roth, P., "Automated testing for electronics manufacturing in-circuit testers", Circuits Manuf., 18 (8), 36-8 (1978).

2929. Lepagnol, J. H., "Automatic soldering techniques", Electron. and Microelectron. Ind., no. 226, 46-9 (1976), French, EEA80-5805.

2930. Livingstone, A. W., "Automatic testing of complex integrated circuits", Post Off Electr. Eng. J, 70 (pt. 3), 161-7 (1977).

2931. Ludwig, D. P., "Chips-in-tape: A study in automated hybrid IC assembly", Electron. Packag. & Prod., 18 (4), 77-86 (1978).

2932. Lyman, J., "Demands of LSI are turning chip makers towards automation, production innovations", Electronics, 50 (15), 81-92 (1977).

2933. Lyman, J., "Techniques of automatic wiring multiply", Electronic, 51 (11), 134-8 (1978), EEA81-32951.

2934. McKee, W. D., Jr. and Unger, R. F., "Hybrid circuit reliability improvement via automation", Proc. of the Tech. Program, Int. Microelectron. Conf., 1977, p. 190-2.

2935. McKeown, P. A., "The place of quality control in automated manufacturing", Qual. Assur., 3 (4), 109-15 (1977), EEA81-13327.

2936. Montante, J. M., Rodrigues de Miranda, W. R. and Oswald, R. G., "Wafer bumping for tape-automated bonding", Proceedings of the 1977 International Microelectronics Symposium, p. 115-19.

2937. Nan, N. and Feuer, M., "A method for the automatic wiring of LSI chips", Proceedings of the 1978 IEEE International Symposium on Circuits and Systems, p. 11-16, EEA81-42174.

2938. Nemeth, P. and Fulop, S., "Automated thermocompression bonding", Finommech. & Mikrotech., 15 (7), 199-201 (1976), Hungarian, EEA80-511.

2939. Oswald, R. G., Montante, J. M. and Rodrigues de Miranda, W. R., "Automated tape carrier bonding for hybrids", Solid State Technol., 21 (3), 39-48 (1978), EEA81-33038.

2940. Pavlidis, T., "A minimum storage boundary tracing algorithm and its application to automatic inspection", IEEE Trans. Syst., Man & Cybern., SMC-8 (1), 66-9 (1978), EEA81-9257.

2941. Pericich, C. E., "Component layout for automatic assembly [proper design cuts assembly costs]", Circuits Manuf., 18 (1), 54, 56, 58 (1978), EEA81-45719.

2942. Preti, S. G., "Use of the minicomputer in the production of hybrid circuits: Automatic hybridising equipment", Electrocompon. Sci. & Technol., 5 (1), 61 (1978), EEA81-37920.

2943. Redemske, R. F., "Technical and economic considerations in selecting automatic wire bonding systems for a custom hybrid house", Proceedings of the 1977 International Microelectronics Symposium, p. 88-92.

2944. Redemske, R. F., "Automatic wire bonding for custom hybrids [economic advantages]", Circuits Manuf., 18 (1), 60, 62-5 (1978), EEA81-45766.

2945. Restrick, P. C., III, "An automatic optical printed circuit inspection system", Proceedings of the Society of Photo-Optical Instrumentation Engineers, Vol. 116, Solid State Imaging Devices, 1977, p. 76-81, EEA81-37892.

2946. Rose, A. S., Scheline, F. E. and Sikina, T. V., "Metallurgical considerations for beam tape assembly [automated bonding and device reliability]", Solid State Technol., 21 (3), 49-52, 68 (1978), EEA81-33039.

2947. Salmon, G. J. E., "Automatic short-circuit testing as a separate check", New Electron., 10 (22), 110, 112, 114 (1977), EEA81-18335.

2948. Schenkel, A. and Rurack, J., "Computer-controlled testing of electronic boards", Bull. Assoc. Suisse Electr., 68 (6), 288-93 (1977), German, EEA80-18151.

2949. Sheets, L., "Computer applications and hybrid microcircuits", Report BDX-613-2020, Bendix Corp., Kansas City, Mo., Department of Energy, 1978, 7 pp.

2950. Shirone, O., Tanabe, T. and Ishizaka, M., "Application of micro-computer to fully-automatic wire-bonder", Natl. Tech. Rep., 23 (6), 1095-1101 (1977), Japanese, EEA81-28184.

2951. Simpson, R. K., "A software package for automatic testing of analog integrated circuits", Digest of Papers 1977 Semiconductor Test Symposium, p. 125-30, EEA81-19527.

2952. Snyder, W. E., "Visual understanding of hybrid circuits via procedural models: an approach to automatic inspection of hybrid circuits", Thesis, Univ. Illinois, Urbana, 1975, 192 pp., Order No. 75-24408, EEA79-13057.

2953. Soffa, A., "Remarks on wire and die bonding for hybrid circuits", Electrocompon. Sci. & Technol., 4 (3-4), 157-61 (1977), EEA81-19169.

2954. Streubel, R., "Automatic digital acquisition of graphical information from design sketches for printed circuit boards", Fernmeldetechnik, 18 (3), 89-90 (1978), German, EEA81-49487.

2955. Thissen, F. L. A. M., "An equipment for automatic optical inspection of connecting-lead patterns for integrated circuits", Philips Tech. Rev., 37 (4), 77-88 (1977), EEA81-22867.

2956. Tsuda, E., Inari, T., Inoue, T., Kosaka, N. and Yamassaki, M., "Development of automated bonding systems for semiconductor devices", Mitsubishi Denki Giho, 51 (11), 730-4 (1977), Japanese, EEA81-33036.

2957. Turner, D. R., Evarts, W. W., Duston, R. W. and Byars, G. L., "Continuous plating of ESS circuit pack receptacle connectors. II. Automatic strip-plating of connector terminals for electronic switching system circuit packs", Western Electric Eng., 22 (2), 32-40 (1978), EEA81-45708.

2958. Voznesensky, S., Domnenko, G., Kondratiuc, I., Malyschenko, Y., Octiabrsky, G., Razdobreyev, A., Stytsyura, L., Tchipulis, V. and Sharshunow, S., "Automated fault location system for digital circuits production", Proceedings of the 6th Triennial World Congress of the International Federation of Automatic Control, Pt. IV, 1975, p. 56.2/1-8, EEA80-9206.

2959. Wolfe, G., "Automated laser trimming with linear motor beam positioning", Circuits Manuf., 15 (10), 46-53 (1975), EEA79-12984.

2960. Yoshizumi, S. and Shimada, Y., "Automatic soldering method by light beam", Natl. Tech. Rep., 23 (2), 323-30 (1977), Japanese, EEA81-9240.

2961. Young, W., "On-site microprocessor-controlled portable module testers", IEEE Trans. Instrum. Meas., IM-27 (2), 147-51 (1978).

## XXV.   EQUIPMENT AND TOOLING

2962. Ames, C., "Wave-solder assembly", IBM Tech. Disclosure Bull., 20 (3), 960-1 (1977), EEA81-18356.

2963. Anon., "Automatic testing: an overview", Circuits Manuf., 16 (2), 17-19, 22 (1976), EEA79-31928.

2964. Anon., "Now – 3A regulation from a TO3 package", Electron. Ind., 2 (5), 17, 19 (1976), EEA79-37029.

2965. Anon., "Precatalyzed LRM system promises lower cost encapsulation", Plast. Eng., 32 (7), 25-6 (1976).

2966. Anon., "Guide to wave soldering equipment", Insul./Circuits, 23 (2), 38-46 (1977).

2967. Anon., "Now you can punch, form, and encapsulate in a new multiple purpose injection mold", Plast. Des. Process, 17 (3), 23-6 (1977).

2968. Anon., "Custom designed ultrasonic cleaning equipment meets stringent cleanliness requirements", Electron. Packag. & Prod., 17 (10), 117-18, 120 (1977).

2969. Anon., "A prototype printed circuit board facility", Report AERE-R9041, UKAEA, Harwell, Oxon., England, 1978, 10 pp., EEA82-614.

2970. Antes, A., "Thick film conveyor furnaces – current status", Solid State Technol., 18 (11), 12 (1975), EEA79-8987.

2971. Arnhart, J. C., Faure, L. H. and Johnson, A. H., "Wire insertion and merging device", IBM Tech. Disclosure Bull., 20 (7), 2673-4 (1977), EEA81-45909.

2972. Bahnck, N., Booth, R. H., Boyer, J. A. and Monahan, J. J., "Methods and apparatus for bonding an article to a substrate", Patent USA 3946931, Publ. March 1976.

2973. Baker, M. R., Noel, Y. J., Shutts, L. and Thomson, W. G., "Apparatus for sealing a semiconductor package", IBM Tech. Disclosure Bull., 18 (8), 2596-9 (1976), EEA79-16906.

2974. Barnes, D. L., Sandwick, T. E. and Van Sickle, R. C., "Logic card retainer and extractor", IBM Tech. Disclosure Bull., 19 (3), 975-6 (1976).

2975. Becker, G., "Specifying and measuring the working capacities of soldering irons and tips - 1, 2", Assem. Eng., 18 (10, 11), 30-3, 26-9 (1975).

2976. Begakis, N., "An inexpensive thick film furnace", Electrocompon. Sci. & Technol., 3 (2), 113-15 (1976), EEA80-496.

2977. Bernard, J. L., "Apparatus for the formation of coating on a substrate", Patent USA 3925187, Publ. December 1975.

2978. Bhattacharya, S. and Sullivan, E. J., "Measurement method for solder creep resolution of less than 5 microns", IBM Tech. Disclosure Bull., 20 (11), 4824-6 (1978).

2979. Bisset, S., "LSI tester gets microprocessors to generate their own test patterns", Electronics, 51 (11), 141-5 (1978).

2980. Bocinski, T. E., Forster, K. R. and Petrozello, J. R., "Multiple test clip", IBM Tech. Disclosure Bull., 20 (12), 5148-9 (1978).

2981. Bonis, S. A., "Bubbles . . . tiny bubbles or pressure and vacuum effects on a hybrid overcoating system", Proceedings of the 1977 International Microelectronics Symposium,, p. 213-22.

2982. Boyer, J. A., "Dispensing components into test sets", Tech. Dig., no. 39, 9-10 (1975), EEA79-1074.

2983. Buggle, R. N., "Guidelines for qualifying and certifying soldering irons", Insul./Circuits, 24 (4), 23-6 (1978), EEA81-41410.

2984. Bush, T. S., "Local memory for a high-speed digital test system", IEEE Trans. Instrum. Meas., IM-26 (3), 217-20 (1977).

2985. Caccoma, G. A., Koestner, J. H., O'Neill, B. C. and Tappen, F. M., "Chip orientor with shrink compensation for gang placement", IBM Tech. Disclosure Bull., 20 (9), 3449-51 (1978).

2986. Caddock, R. E., "Apparatus for manufacturing cylindrical resistors for thick-film silk-screening", Patent USA 4075968, Publ. February 1978.

2987. Cadwallader, R. H., Darves-Bornoz, Y., Gasparri, A. S., et al, "Apparatus for the printing of ceramic green sheets", Patent USA 4068994, Publ. January 1978.

2988. Carpenter, R. W. and Zucconi, T. D., "Controlled filtering system for chemical solutions", IBM Tech. Disclosure Bull., 20 (1), 72 (1977), EEA81-14277.

2989. Castellani, E. E., Power, J. V. and Romankiw, L. T., "Nickel iron (80 to 20) alloy thin film electroplating method and electrochemical treatment and plating apparatus", Patent USA 4102756, Publ. July 1978.

2990. Chernobrovkin, D. I., Mishanin, N. D., Piganov, M. N. and Goncharov, A. G., "Instrument for matching thin film elements of microcircuits", Prib. & Sist. Upr., no. 6, 45-6 (1978), Russian, EEA81-49499.

2991. Cistola, A. B. and Darrow, R. E., "Customized solder reflow apparatus", IBM Tech. Disclosure Bull., 20 (7), 2612 (1977), EEA81-45768.

2992. Craft, W. H., "Silicon etching machine for beam-lead IC fabrication", Circuits Manuf., 17 (9), 44-50 (1977).
2993. Daebler, D. H. and Malmgren, R. P., "Apparatus for repairing hybrid circuits", Patent USA 3904100, Publ. September 1975.
2994. Damm, E. P., Jr., "Convection cooling apparatus", IBM Tech. Disclosure Bull., 20 (7), 2755-6 (1977).
2995. Dehaine, G. and Kurzweil, K., "Tape automated bonding moving into production", Solid State Technol., 18 (10), 46-52 (1975), EEA79-4783.
2996. Denslow, C. A., "Ultrasonic soldering equipment for aluminum heat exchangers", Weld J., 55 (2), 101-7 (1976).
2997. Doak, K. W., "A primer on vacuum pumps", Electron. Packag. & Prod., 17 (7), 145-8 (1977).
2998. Doo, V. Y. and Johnson, A. H., "Air board replenishing system", IBM Tech. Disclosure Bull., 20 (11), 4341-2 (1978).
2999. Dubey, G. C., "The squeegee in printing of electronic circuits", Microelectron. & Reliab., 14 (5-6), 427-9 (1975), EEA79-24895.
3000. Dugas, R. A., "Thermal bonder heater sensor", Tech. Dig., no. 42, 9-10 (1976), EEA79-33165.
3001. Ett, A. H., "Tool for handling, inserting and soldering MOS logic in dip packages", IBM Tech. Disclosure Bull., 21 (1), 389-90 (1978).
3002. Firdaus, A. and Scacciaferro, F., "Automatic height sensing and adjusting system", IBM Tech. Disclosure Bull., 19 (9), 3502-3 (1977), EEA80-35956.
3003. Foytlin, L. F., "Wire cleaning tool with take-up improves winding and encapsulation", Insul./Circuits, 22 (9), 19-20 (1976), EEA80-37.
3004. Fulop, S. and Nemeth, P., "Assembly problems of integrated circuits (automated bonding)", Finommech. & Mikrotech., 14 (9), 261-4 (1975), Hungarian, EEA79-9008.
3005. Galinskii, E. R., Sarandi, R. L., Khion, P. V. and Khutoryanskii, E. D., "UDS-3M equipment for diffusion bonding", Weld. Prod., 23 (2), 63-4 (1976), EEA80-8590.
3006. Ganopolskii, L. S., Ryabinin, V. I. and Tsypin, B. V., "An assembly control tester for radio-electronic equipment and hybrid integrated circuits", Prib. & Sist. Upr., no. 1, 45-6 (1978), Russian, EEA81-22825.
3007. Garnett, L. T., "Zero maintenance pressure instrumentation", Offshore Technol. Conf. 10th Annu., Proc., Vol. 2, 1978, p. 671-9.
3008. Gasparri, A. S. and Zykoff, F. B., "Via hole inspection apparatus", IBM Tech. Disclosure Bull., 20 (4), 1414 (1977), EEA81-22857.
3009. Gisler, H. J., Jr., "G.E. extends automation to thick-film hybrid production", Electron. Packag. & Prod., 17 (11), 105-10 (1977).
3010. Harrod, R. D. and Moore, H. R., "Ungated common I/0 buffer for card testing", IBM Tech. Disclosure Bull., 21 (6), 2476-7 (1978).

3011. Hester, R. L., Kobesky, L. J., Kriger, B. J. and Lawrence, R. W., "Module tester", IBM Tech. Disclosure Bull., 21 (6), 2277-8 (1978).

3012. Hieke, E. and Meusburger, G., "SEM stroboscopy for the evaluation of IC operating functions in the subnanosecond range", Rev. Sci. Instrum., 49 (6), 802-5 (1978).

3013. Hoffman, K. M., "Device removal tool for metallized ceramic substrates", IBM Tech. Disclosure Bull., 21 (3), 959-60 (1978).

3014. Jose, D. B., "Flexible, general-purpose tester for incoming inspection of custom-integrated circuits", SME Tech. Pap. Ser. EE, Paper EE77-729, 1977, 14 pp.

3015. Kan, A. G. and Nikitin, A. G., "An instrument for testing parameter deviations of microelectronic devices", Poluprovodn. Tekh. & Mikroelektron., no. 24, 96-8 (1976), Russian, EEA80-489.

3016. Kashioka, S., Ejiri, M. and Sakamoto, Y., "A transistor wire-bonding system utilising multiple local pattern-matching techniques", IEEE Trans. Syst. Man. & Cybern., SMC-6 (8), 562-70 (1976), EEA79-31939.

3017. Kazakov, N. F., Sergeev, A. V. and Lakin, N. A., "Devices for making precise measurements of stresses and deformation in diffusion welding", Autom. Weld., 29 (7), 52 (1976), EEA80-23996.

3018. Kehagioglou, T., "Mirrored alignment gauge", IBM Tech. Disclosure Bull., 21 (5), 1962-3 (1978).

3019. Keizer, A. and Brown, D., "Bonding system for microinterconnect tape technology", Solid State Technol., 21 (3), 59-64 (1978), EEA81-33041.

3020. Kimura, S. and Hojo, T., "Automatic bonding machine - a new assembly technique for reducing production cost", JEE J. Electron. Eng., no. 98, 44-6 (1975).

3021. Kotrch, G. S., "Snap-in substrate holder for coil winder", IBM Tech. Disclosure Bull., 20 (9), 3411-2 (1978).

3022. Kovac, M. G., Chleck, D. and Goodman, P., "New moisture sensor for in-situ monitoring of sealed packages", Solid State Technol., 21 (2), 35-9, 53 (1978).

3023. Kutch, G., "Inspection apparatus for apertured green sheets", IBM Tech. Disclosure Bull., 20 (7), 2678-9 (1977), EEA81-45192.

3024. Leiseder, L., "Electronics production - plant and equipment", Tech. Rundsch., 68 (47), 5, 7 (1976), German, EEA80-9189.

3025. Linsley, J. W. and Porod, R. F., "Capacitor package and assembling fixture", Tech. Dig., no. 39, 23-4 (1975), EEA79-88.

3026. Lyman, J., "Automated circuit testers lead the way out of continuity maze", Electronics, 46 (16), 87-95 (1975).

3027. Mansfeld, F., "Copper plating bath monitor", Plat. Surf. Finish., 65 (5), 60-2 (1978).

3028. Markstein, H. W., "Vacuum deposition equipment trends", Electron. Packag. & Prod., 17 (9), Pt. 1, 36-8, 40, 42, 44 (1977).

3029. Martens, D., "Automatic test systems – in form of mass produced or purpose built equipment?", Elektron. Ind., 7 (3),
56 (1976), German, EEA79-28607.
3030. Meeks, A., "Fluidic card reader", IBM Tech. Disclosure Bull.,
21 (5), 1964-5 (1978).
3031. Meeks, A. E., "Solder reflow apparatus", IBM Tech. Disclosure
Bull., 21 (7), 2918-19 (1978).
3032. Megivern, C. F., "Digital delay technique", IBM Tech. Disclosure Bull., 21 (7), 2794-5 (1978).
3033. Meyer, D. E., "Miniature moisture sensor for in-package use
by the microelectronics industry", 13th Annual Proceedings
of Reliability Physics Symposium, 1975, p. 48-52, EEA79-48865.
3034. Moran, J. M. and Saunder, T. E., "High precision laser machining circuit generator", Rev. Sci. Instrum., 46 (9),
1267-72 (1975), EEA79-8972.
3035. Moran, K. P., Pascuzzo, A. L. and Yacavonis, R. A., "Spring-
loaded module connectors for mounting an array of modules
on circuit board", IBM Tech. Disclosure Bull., 20 (9), 3434-5
(1978).
3036. Mracek, J., "Methods and apparatus for selectively removing
a metallic film from a metallized substrate", Patent USA
4081654, Publ. March 1978.
3037. Neundorf, H., "Experimental connecting boxes for integrated
circuit experiments", Radio Fernsehen Elektron., 25 (8), 260-2
(1976), German, EEA79-33088.
3038. Nowakowski, A. and Jackowski, J., "Temperature measuring
device for thick-film IC's", Elektronika, 16 (7-8), 312-14
(1975), Polish, EEA79-1028.
3039. Odawara, G. and Ishiwata, S., "Automated wire bonder", J.
Jap. Soc. Precis. Eng., 42 (10), 948-54 (1976), Japanese,
EEA80-9207.
3040. Onufer, R. J. and Stanley, E. C., "Using a rotational visco-
meter to characterize thick film materials. I.", Insul./Cir-
cuits, 21 (13), 31-6 (1975), EEA79-13003.
3041. Onufer, R. J. and Stanley, E. C., "Using a rotational visco-
meter to characterize thick film materaials. II.", Insul./
Circuits, 22 (1), 23-7 (1976), EEA79-16867.
3042. Pavlov, D. K., Vangelov, V. K. and Popov, K. M., "Transformer
transport and epoxy resin encapsulation equipment", Elektro
Prom.-st. & Priborostr., 10 (10), 425-6 (1975), Bulgarian,
EEA79-19569.
3043. Pedrotti, D. G., "Automatic wire bonding of IC's", International Microelectronics Conference, Proceedings of the Technical Program, 1975, p. 125-8.
3044. Petrov, A. I., "Light spot projector system for semiautomatic
chip assembly machine", Elektro Prom.st.- & Priborostr., 13
(1), 29 (1978), Bulgarian, EEA81-33034.

3045. Popescu, I. M., Stanciu, G. A. and Stoichita, C. M., "A new laser system for testing semiconductor devices and ICs", Bul. Inst. Politeh. 'Gheorghe Gheorghiu-Dej' Bucuresti Ser. Electroteh., 39 (4), 11-15 (1977), Rumanian, EEA81-45962.

3046. Radzik, E. P., "Circuit net detector module", IBM Tech. Disclosure Bull., 19 (8), 3123-4 (1977), EEA80-32233.

3047. Ramey, D., "Thick-film screen printing equipment survey", Electron. Packag. & Prod., 17 (9), Pt. 1, 49-54 (1977).

3048. Ring, D. A., Sitler, W. R., Symons, M. E. and Youngs, L. J., "Data processor diagnostic test system", IBM Tech. Disclosure Bull., 21 (1), 115-17 (1978).

3049. Ripka, G. and Albrecht, M., "A digital program to control drawing machines in preparing master patterns for thick-film integrated circuits", Finommech. & Mikroteh., 15 (7), 193-8 (1976), Hungarian, EEA80-497.

3050. Ross, T., "Optical scanning system for defect detection", IBM Tech. Disclosure Bull., 20 (9), 3431 (1978), EEA82-118.

3051. Runyon, S., "Focus on IC testers", Electron. Des., 24 (9), 48-55 (1976), EEA79-31931.

3052. Ryan, W. J., Schiller, H. H. and Zykoff, F. B., "Punch apparatus", IBM Tech. Disclosure Bull., 21 (2), 554-5 (1978).

3053. Schaefer, T. L., "Vacuum-assisted hand-guided test probe", IBM Tech. Disclosure Bull., 20 (7), 2792-3 (1977), EEA81-45727.

3054. Schott, F. A., "Boat thermal enhancement for semiconductor chips and modules", IBM Tech. Disclosure Bull., 20 (7), 2635 (1977).

3055. Sheeley, J. D., "Development of an improved bonding fixture", Report BDX-613-1252, Bendix Corp., Kansas City, Mo., 1976, 21 pp.

3056. Smolski, R. and Roman, W., "Holders for hybrid integrated circuit substrates", Elektronika, 16 (6), 257-60 (1975), Polish, EEA79-1102.

3057. Strejcek, P. and Zakopal, J., "The Multiset 24 - a fully programmable meter of integrated circuits", Sdelovaci Tech., 23 (7), 255-8 (1975), Czech, EEA79-4722.

3058. Sulyok, J., "Cryptoclimate of electric equipment in the case of wet climate surroundings", Finommech. & Mikrotech., 15 (5), 136-41 (1976), Hungarian, EEA79-44776.

3059. Tomic, P., "Failure analysis equipment and services", Solid State Technol., 21 (10), 74-6 (1978).

3060. Varmazis, C., Viswanathan, R. and Caton, R., "Technique for bonding gold and silver metals on sapphire", Rev. Sci. Instrum., 49, 549-50 (1978), CA88-181260.

3061. Venkin, G. V., Kulyuk, L. L. and Maleev, D. L., "Circuit for precision thermostating based on an integrated microcircuit", Instrum. & Exp. Tech., 18 (4), Pt. 2, 1286-7 (1975), EEA79-24878.

3062. Vojtechnovsky, K., "Laser microellipsometers LEM-2 and LEF-M", Cesk. Cas. Fis. Sekce A, 27 (1), 75-6 (1977), Czech, EEA80-28619.

3063. Walther, V., Urban, H., Werner, V., Horn, J. and Neukirchner, W., "Drying station of the washing-drying equipment WTA 1 for the manufacturing of printed circuit boards", Feingeratetechnik, 27 (7), 317-20 (1978), German, EEA81-49493.

3064. Williams, E. E., "Apparatus for coating substrates", Patent USA 4010710, Publ. March 1977.

3065. Wojcik, Z. M., "The application of automatic objects centering in microelectronics", Elektronika, 17 (11), 409-11 (1976), Polish, EEA80-20575.

3066. Yoshida, S., "Optical instruments for manufacturing integrated circuits", Oyo Buturi, 44 (5), 526-32 (1975), Japanese, EEA79-8971.

3067. Young, G. H., "Automatic machines for the manufacture of hybrid integrated circuits", Autom. & Inf. Ind., no. 56, 47-51 (1977), French, EEA80-28631.

## XXVI.   COST AND YIELD

3068. Acello, S., "Mini-Pak: a cost-effective leadless chip carrier", Electron. Packag. & Prod., 17 (6), 78-82 (1977), EEA81-14309.

3069. Anon., "Low cost encapsulation process", Polym. News, 3 (2), 91-2 (1976), EEA80-31397.

3070. Anon., "Low cost substrate for thick film circuits using porcelain", New Electron., 11 (3), 36 (1978), EEA81-23633.

3071. Anon., "Changing economics in custom ICs", Electron. Eng., 50 (607), 65, 67-8 (1978), EEA81-37894.

3072. Anon., "Cost effective plating", Eng. Mater. & Des., 22 (6), 72 (1978), EEA81-41409.

3073. Bach, W., "Semiconductor circuit elements testing from the point of view of economy", Bauelem. Elektrotech., 11 (9), 62-70 (1976), German, EEA79-46407.

3074. Bernard, J., "IC yield problem: a tentative analysis for MOS/SOS circuits", IEEE Trans. Electron. Devices, ED-25 (8), 939-44 (1978).

3075. Boswell, D. and Campbell, D. S., "Economics of thick and thin film hybrid production in Europe", Electrocompon. Sci. & Technol., 4 (3-4), 219-23 (1977), EEA81-19165.

3076. Crossley, A., "Low cost metal film resistors", New Electron., 11 (5), 128-9 (1978), EEA81-28065.

3077. Dellacqua, R. and Forlani, F., "Organization of hybrid thick film manufacturing for diversified market", International Conference on Manufacturing and Packaging Techniques for Hybrid Circuits, 1976, p. 163-72, French, EEA79-46567.

3078. Devitt, D. and George, J., "Beam tape plus automated handling cuts IC manufacturing costs", Electronics, 51 (14), 116-19 (1978).

3079. Eremin, S. A., Ivantsov, V. V. and Kudinov, V. V., "Estimation of yield probability of operational large-scale integrated circuits with a limited number of faulty components", Izv. VUZ Radioelektron., 21 (1), 83-7 (1978), Russian, EEA81-28171.

3080. Evans, D., "Industry's thick film needs", New Electron., 10 (8), 42 (1977), EEA80-41109.

3081. Gedney, R. W. and Werbizky, G. G., "Low cost integrated circuit", IBM Tech. Disclosure Bull., 20 (9), 3399-400 (1978).

3082. Grossman, M., "Better materials and equipment trim costs and raise reliability", Electron. Des., 23 (15), 24, 26, 28, 30, 32 (1975), EEA79-1101.

3083. Heap, B. C. and France, S. A., "Improved hybrid circuit assembly yields and reliability by glassivation of the semiconductor chip", Electrocompon. Sci. & Technol., 4 (3-4), 117-24 (1977), EEA81-14334.

3084. Hilson, D. G. and Johnson, G. W., "New materials for low cost thick film circuits", Solid State Technol., 20 (10), 49-54, 75 (1977).

3085. Holloway, P. and Norton, M., "A high yield, second generation 10-bit monolithic DAC", 1976 IEEE International Solid-State Circuits Conference, Digest of Technical Papers, p. 106-7, 236, EEA79-46463.

3086. Joyce, B. T., "Hybrids - a look at the total cost", IEEE Trans. Manuf. Technol., MFT-6 (4), 69-72 (1977), EEA81-16477.

3087. Khadpe, S., "Chip carriers, LIDs and SOTs - an analysis of cost and performance tradeoffs for hybrid applications", 27th Electronic Components Conference, 1977, p. 30-3, EEA80-35944.

3088. Koury, R., "When custom LSI can save you money", Mach. Des., 49 (18), 91-5 (1977).

3089. Kronert, R., "Cutting costs for thick film circuits", Elektronik, 27 (8), 64-6 (1978), German, EEA81-45760.

3090. Lassus, M., "Semi-conductors provided with protuberances offer an economic solution for hybrid assemblies", Electron. & Microelectron. Ind., no. 229, 23-5 (1976), French, EEA80-9201.

3091. Lassus, M., "The solder-head method for cost reduction in semiconductor manufacture", Elektron. Anz., 9 (9), 36-9 (1977), German, EEA81-19168.

3092. Ledoux, A. C., "Cost effectiveness of adapting existing military documents for microelectronics to microwave integrated circuits (MIC's)", 12th Electrical/Electronics Insulation Conference, 1975, p. 103-5, EEA79-46554.

3093. Luettge, H., "Low cost protection for active hybrid components", Can. Electron. Eng., 20 (2), 21-2 (1976), EEA79-19566.

3094. Lyman, J., "Process adaptation shrinks interconnection costs", Electronics, 49 (22), 116-18, 21 (1976), EEA80-39.

3095. Miles, T. J., Langlais, D. G. and Daniel, R. P., "Hybrid circuit qualification at reduced cost", Proceedings of the 1977 International Microelectronics Symposium, p. 138-45.

3096. Pinner, F. R., "Leadless device reflow soldering is reliable and lowers assembly costs", Assem. Eng., 20 (10), 50-4 (1977).

3097. Polednak, S. and Wallace, A., "Techniques for yield improvement in complex multi-chip hybrids", Proceedings of the 1977 International Microelectronics Symposium, p. 127-30.

3098. Redditt, J., "Minimizing set-up costs for ATE", Electron.
      Packag. & Prod., 17 (7), 283-7 (1977).
3099. Riemer, D. E., "Cost efficiency of thick-film conductors",
      Solid State Technol., 18 (10), 42-5 (1975), EEA79-4736.
3100. Rubin, W., "Cutting solder costs", Electron. Prod. Methods &
      Equip., 6 (3), 30, 32 (1977), EEA81-13288.
3101. Salzer, J. M., "Cost of multichip assemblies", International
      Microelectronics Conference, Proceedings of the Technical
      Program, 1975, p. 25-9.
3102. Thurman, M., "Cutting production costs with in-circuit test
      systems", Solid State Technol., 21 (10), 77-9 (1978).
3103. Surendran, K. and Ramesh, C. K., "Cost reliability trade off
      in an electronic module", Microelectron. & Reliab., 15 (5),
      493-6 (1976).
3104. VanHise, J. A., "Mesa multilayer ceramic substrate construc-
      tion", IBM Tech. Disclosure Bull., 21 (1), 139-41 (1978).
3105. Yen, J. C., "A high yield and low cost process for building
      multilayer metal structure by using PYRE-M.L.", Electrochem-
      ical Society Fall Meeting, Extended Abstracts, 1975, p. 444-5,
      EEA79-20553.

## XXVII.  FUTURE TRENDS

3106. Aigrain, P. R., "The emerging role of the European IC indus-
      try", 1976 IEEE International Solid-State Circuits Conference,
      Digest of Technical Papers, p. 49, EEA79-46363.
3107. Andrejasich, R. J., "Keyboards as a packaging element", Elec-
      tron. Packag. & Prod., 18 (2), 73-5, 77-8, 80, 82, 84, 86
      (1978), EEA82-103.
3108. Anon., "Coping with the new trends in laminating for flexible
      packaging", Mod. Plast., 54 (12), 40-1 (1977).
3109. Bensieck, H. J., "PTC resistors are self-regulating heating
      elements and as indestructible overload fuses", Bauelem.
      Elektrotech., 11 (12), 41-2, 44 (1976), German, EEA80-13195.
3110. Buchbinder, H. G., "Japanese PC industry report: the best is
      yet to come", Circuits Manuf., 17 (9), 24-36 (1977).
3111. Chevalier, J. G., Eisenhart, R. K. and Harrod, W. L., "Pack-
      aging electronic circuits for the future", Bell Lab. Rec.,
      54 (2), 34-8 (1976).
3112. Ciccio, J. A. and Thun, R. E., "Ultra-high density VLSI
      modules", IEEE Trans. Components, Hybrids & Manuf. Technol.,
      CHMT-1 (3), 242-8 (1978), EEA81-49504.
3113. Cozzens, W. B., "Bright future for laser trimming", Electron.
      Engineering, 48 (576), 58-9 (1976), EEA79-11908.
3114. Crossley, A., "The future of thin and thick film metal resis-
      tors", New Electron., 10 (10), 82, 84, 86, 89 (1977), EEA81-5534.
3115. Del Rosso, V., "Future trends in electronically controlled
      machine functions", Package Dev. Syst., 7 (3), 21-5 (1977).
3116. Erlandson, P. M., "New concepts in packaging: Metal cans of
      the future", Package Dev. Syst., 7 (2), 31-7 (1977).

3117. Gedney, R. W., "Trends in packaging technology", 16th Annual
      Proceedings Reliability Physics Symposium, 1978, p. 127-31,
      EEA81-48756.
3118. Hagstrom, R. A., "Hybrid technology's real potential is only
      now emerging", Electron. Des., 26 (13), 102-5 (1978).
3119. Himmel, R. P., "A new generation of large hybrid modules -
      SLIM", Solid State Technol., 20 (10), 68-73 (1977).
3120. Hirsl, J., "New tendencies in the production of hybrid inte-
      grated circuits", Sdelovaci Tech., 25 (9), 335-6 (1977),
      Czech, EEA81-10056.
3121. Jardine, L. J., "Trends in high-density connector/intercon-
      nection technology", Electron. Packag. & Prod., 18 (10),
      47-50, 52, 54, 56-7 (1978).
3122. Kutsenko, V. I., Kolesnikov, N. D. and Shabelnik, N. M.,
      "Basic trends in the development of welding and soldering in
      low-voltage equipment manufacturing", Sov. Electr. Eng., 47
      (1), 47-50 (1976), EEA80-24000.
3123. Lyman, J., "Packaging technology responds to the demands for
      higher densities", Electronics, 51 (20), 117-25 (1978).
3124. Marathe, B. R., "Recent trends in thick-film hybrid circuits",
      J. Inst. Electron. Telecommun. Eng., 24 (5), 213-18 (1978).
3125. Markstein, H. W., "Fiber optics for systems packaging",
      Electron. Packag. & Prod., 18 (10), 60-2, 64, 66-7 (1978).
3126. Matcovich, T. J., "New chip carrier package concepts offer
      unprecedented electronic system flexibility", 27th Electronic
      Components Conference, 1977, p. 16-24, EEA80-35959.
3127. Minarsky, E., "The trends in thick film technology", Sdelovaci
      Tech., 25 (8), 308-9 (1977), Slovak, EEA81-5625.
3128. Mohri, K., Shimano, M., Kuroyanagi, T. and Oguino, M., "Chroma
      systems trend in the past and future and latest chroma IC with
      versatile flexibility", IEEE Trans. Consum. Electron., CE-24
      (1), 81-8 (1978).
3129. Mones, A. H. and Rosenberg, R. M., "Trends in thick film
      materials", Solid State Technol., 19 (10), 47-9 (1976),
      EEA80-3300.
3130. Nagy, J., "RF connector trend: smaller and better", Electron.
      Packag. & Prod., 17 (5), C27-C31 (1977).
3131. Pfeifer, H. J. and Wetzko, M., "Layer hybrid technology - a
      modern development trend of microelectronics. II.", Technik,
      32 (6), 317-19 (1977), German, EEA81-10057.
3132. Pircher, G., "Microlithography: general principles, uses and
      trends", Rev. Tech. Thomson-CSF, 10 (1), 5-43 (1978), French,
      EEA81-49497.
3133. Seltzer, R., "Future test: how to cope with the changing world
      of LSI", Circuits Manuf., 17 (12), 33-5, 38, 41 (1977).
3134. Sergent, J. E., "Trends in thick-film materials", Electron.
      Packag. & Prod., 18 (9), 39-40, 42, 44 (1978).
3135. Siewiorek, D. P., Thomas, D. E. and Scharfetter, D. L., "Use
      of LSI modules in computer structures trends and limitations",
      Computer, 11 (7), 16-25 (1978).

3136. Silver, N., "Hybrids: the state of the art and some future trends", Electron, no. 103, 60, 64 (1976), EEA80-503.
3137. Stein, S. J., Spadafora, L. and Huang, C., "New developments in thick film conductors", Solid State Technol., 22 (1), 74-80 (1979).
3138. Tsunashima, E., "Research injects new creativity into high-density packaging", JEE, no. 104, 26-30 (1975), EEA79-4362.
3139. Umbaugh, C. W., "New packaging technology for Honeywell large scale computer system", 14th IEEE Computer Society International Conference, 1977, p. 263-6, EEA81-5639.
3140. Wuich, W., "Present position and development trends in welding technology", Technica, 27 (7), 431-4 (1978), German, EEA81-27196.
3141. Zinschlag, H. P., Conover, J. A. and Long, W. E., "Potential impact of LSI on process control systems", Advances in Instrumentation, Vol. 30, Pt. I, 1976, p. 508/1-5, EEA79-46384.